Energy Medicine for Women

Other books coauthored by Donna Eden and David Feinstein

Energy Medicine
The Promise of Energy Psychology

JEREMY P. TARCHER/ PENGUIN

a member of

Penguin Group (USA) Inc.

New York

Energy Medicine for Women

*Aligning Your Body's Energies to Boost
Your Health and Vitality*

Donna Eden

with David Feinstein, Ph.D.

Photography by Christine Alicino

JEREMY P. TARCHER/PENGUIN
Published by the Penguin Group
Penguin Group (USA) Inc., 375 Hudson Street, New York, New York 10014, USA •
Penguin Group (Canada), 90 Eglinton Avenue East, Suite 700, Toronto, Ontario M4P 2Y3, Canada
(a division of Pearson Canada Inc.) • Penguin Books Ltd, 80 Strand, London WC2R 0RL, England •
Penguin Ireland, 25 St Stephen's Green, Dublin 2, Ireland (a division of Penguin Books Ltd) •
Penguin Group (Australia), 250 Camberwell Road, Camberwell, Victoria 3124, Australia
(a division of Pearson Australia Group Pty Ltd) • Penguin Books India Pvt Ltd, 11 Community Centre,
Panchsheel Park, New Delhi–110 017, India • Penguin Group (NZ), 67 Apollo Drive, Rosedale,
North Shore 0632, New Zealand (a division of Pearson New Zealand Ltd) • Penguin Books (South Africa)
(Pty) Ltd, 24 Sturdee Avenue, Rosebank, Johannesburg 2196, South Africa

Penguin Books Ltd, Registered Offices:
80 Strand, London WC2R 0RL, England

Most Tarcher/Penguin books are available at special quantity discounts for bulk purchase for sales promotions, premiums, fund-raising, and educational needs. Special books or book excerpts also can be created to fit specific needs. For details, write Penguin Group (USA) Inc. Special Markets, 375 Hudson Street, New York, NY 10014.

Library of Congress Cataloging-in-Publication Data

Eden, Donna.
Energy medicine for women : aligning your body's energies to boost your health and vitality / Donna Eden with David Feinstein.
p.   cm.
Includes bibliographical references and index.
ISBN 978-1-58542-647-8
1. Energy medicine.   2. Women—Health and hygiene.   I. Feinstein, David.   II. Title.
RZ421.E34   2008   2008018637
613'.04244—dc22

Printed in the United States of America
7   9   10   8   6

Neither the publisher nor the authors are engaged in rendering professional advice or services to the individual reader. The ideas, procedures, and suggestions contained in this book are not intended as a substitute for consulting with your physician. All matters regarding your health require medical supervision. Neither the authors nor the publisher shall be liable or responsible for any loss or damage allegedly arising from any information or suggestion in this book.

While the authors have made every effort to provide accurate telephone numbers and Internet addresses at the time of publication, neither the publisher nor the authors assume any responsibility for errors, or for changes that occur after publication. Further, the publisher does not have any control over and does not assume any responsibility for author or third-party websites or their content.

Dedicated to my daughters,
Tanya Dahlin and Dondi Dahlin—
may they and their generation soar!

Jeremy P. Tarcher, who has been the trumpet for hundreds of our culture's most inspiring voices, has brought special notice to my own voice as well. I am grateful for his encouragement, his wisdom, and his pioneering spirit.

Sara Carder at Tarcher/Penguin has been a dream of an editor.

The late Jeffrey Harris, M.D., was a wise and wonderful source of guidance and support to me throughout this book.

Every person involved in our growing energy medicine teaching community has contributed in his or her own special way to the development of the ideas presented here.

Credits

Photographs **by Christine Alicino**

Illustration Models
Christine Alicino
Infant Audrey Dahle
Cindy Dahle
Dondi Dahlin
Tanya Dahlin
Donna Eden
Rose Harris
Gwen Mazur
Catherine Potenza
Beate Priolo
Tia Via

Contents

There is a vitality, a life force, an energy, a quickening,
that is translated through you into action, and because
there is only one of you in all time, this expression
is unique. And if you block it, it will never exist
through any other medium and will be lost.

—MARTHA GRAHAM

Foreword

Christiane Northrup, M.D.

began experimenting with energy medicine in my gynecological practice many years ago, using it when I had to perform invasive office procedures such as endometrial biopsies, the removal of IUDs, or tests that involved injecting dye into the uterus. I sensed how disruptive these routine procedures were to the body's natural energy fields, so as the final part of the treatment, I would have my patient lie down and I would move my hands in long passes above her body. Something in this simple act seemed to help restabilize the body's energy fields. Many women would report immediate relief of pain or cramps, pleasurable tingling sensations in the areas that had been traumatized, or a deepening calm throughout their bodies.

While about as simple a technique as exists, it was pleasant, noninvasive, and effective. Now, even as the field of hands-on energy medicine has developed far more sophisticated procedures than my early experiments, the approach has retained these basic qualities: simple, pleasant, noninvasive. But don't be fooled. Your hands hold powerful medicine—"energy medicine"—that beyond just soothing the body after an invasive procedure is able in itself to prevent or help overcome challenging illnesses.

Energy medicine is moving into our health care system quickly, powerfully, and none too soon. We have as a society come to the end of our journey in Newtonian medicine, a perspective that looks at the body more like a bag of organs and bones than a miracle of animation; that focuses on illness rather than optimizing health; and that often futilely seeks to identify simple cause-and-effect relationships rather than to grasp how body, mind, and spirit are profoundly interrelated. We have as a result built a medical system based on drugs and surgery that usually doesn't become meaningfully involved in health care until after the person is already sick. We are like a river patrol that sends powerboats out into the rapids to rescue people who are drowning rather than going upstream and figuring out how to keep them from falling into the rapids in the first place.

We can't go much further with this model. It isn't working, as Michael Moore's documentary *Sicko* makes abundantly and compassionately clear. The old paradigm is breaking down before our eyes. But we also see the emergence of a new paradigm and, happily, one whose ancient roots have withstood the test of time. The new paradigm, as this book beautifully articulates, addresses biological processes at their energetic foundation; gives rise to methods that are precise, practical, rapid, and noninvasive; optimizes health as well as countering illness; empowers the person with effective methods for back-home self-care; and integrates body, mind, and spirit.

But perhaps the best news about energy medicine is that even though many leading-edge thinkers in health care believe it to be the medicine of the future, it is available today. You don't have to wait. So what if it's not yet in your town! So what if it's not yet in your hospital! So what if your doctor doesn't yet believe in it! You can learn the basics from the book you are reading. And you can use it with any medical regimen you are currently on. Any regimen! A beauty of energy medicine is that it won't interfere. There are no side effects whatsoever. So why wouldn't you try it? There is nothing to lose!

This book is a wonderful introduction to energy medicine. It distills the experiences of one of the field's most joyous and effective pioneers into a heartful, user-friendly, enormously practical guide. I agreed to write this foreword before I met Donna Eden or had a chance to see her work. While I review many well-written books that provide good information, there is a deeper ring of truth that is precious and rare. This book sang with it. When I finally had the pleasure of meeting Donna, it was evident that the compassion and wisdom I felt in her book reflect the integrity in the being of its author. Incredibly youthful in her sixty-fifth year, Donna is her own best advertisement. She is ebullient with energy. And she is very, very fun to be around.

Writing this foreword is part of a full circle in my relationship with energy medicine. I have always recognized the importance of the body's energies in my medical practice. I intuitively understood the influence of my patients' thoughts, fears, desires, relationships, family history, jobs, diet, use of exercise, and overall lifestyle on their body's energies and the powerful impact of those energies on their health and their illnesses. But this fundamental dimension of health care isn't taught in medical schools, and as I began to speak with my colleagues about its importance, I felt like the proverbial voice in the wilderness. In the early part of my career, I didn't think this perspective would be accepted during my lifetime. But now, with the embrace of these ideas reflected in many arenas, including in my own life—from having my books on the *New York Times* best-seller list to multiple appearances on *Oprah*—I have clear-cut, measurable evidence that a broad population is ready to hear about the principles that have guided my work. And the medical community is listening as well.

The world is in fact changing faster than most of us imagined could be possible even a decade ago. While many of the changes are shaking us down to our foundations, a hopeful landscape is also emerging. Energy medicine is one of its contours. It is not just the latest fad in alternative health care. It is a fundamentally new way of grasping who we are. During this precarious time in our history, as you will read in Donna's introduction, "a book about the energies that animate a woman's body is obliged to mainline you into a deeper connection with the archetypal feminine principle that must be re-embraced if our species is to survive. . . . The archetypal feminine is not just a set of ideas and values somewhere out in the ethers. It is coded, yes, in your energies, but also in your genes, your hormones, and the actions they cause you to take. [At their core] our feminine instincts are toward love, cooperation, justice, compassion, family, nature, and peace." Energy medicine brings these values into health care, and it is not just for women or just for healers. Every mother, father, teacher, police officer, and politician needs to become a quick study in energy medicine to better cope with the challenges we all face in today's world.

Most women, however, are innately sensitive to energy in a way that is different from most men. We are more multimodal in the ways our brains are constructed. We are also the healers of our families. When women are healthy and happy, their families are healthy and happy. So you aren't reading this book just for yourself. You are also reading it for the better mother, wife, sister, and leader you will become.

I want to emphasize and comment on a few of my favorite concepts from the book. It is absolutely lovely the way Donna conveys that when she is with a client, she starts by seeing beyond the person's energy blockages to the radiant soul that is inside each of us. This leads to health care where there is no judgment, no blame, no shaming, no guilt, no sense of doing it wrong. There is simply the instruction: "This is how you heal." When your healer sees you this truly and deeply, you experience the radiant self that is within you. The book gives you a vicarious sense that attunes you to your own radiant self.

A second major concept in this book is the basic premise of energy medicine, which is that Einstein was right—*energy is all there is*. The physical world is nothing but a slower vibration of energy. Energy is the prime mover of all we see and know. This understanding opens you to a new way of taking charge of your health. You change the energy and, over time, your body *has to* respond. This is an enormously empowering idea. It flies in the face of conventional wisdom, where we believe that treating illness requires something from the outside, such as a drug or a surgical procedure. Energy healing is an inside job. As you begin to do the exercises in this book, you will feel a difference as the energy begins to shift within you. But you don't need to believe me. See for yourself. As you start to use these methods, you will have irrefutable evidence that they work. And from these experiences, you will develop a rock-solid knowledge about the central role of your body's energies. I was impressed when Donna told me about being alerted to a discussion on a cystic fibrosis blog. Double-lung-transplant patients were figuring out together how to use energy techniques (based on Donna's first book, *Energy Medicine*) and sharing their successes. So no matter what the concerns that caused you to pick up this book—whether PMS, menstrual cramps, infertility, premature ovarian failure, or menopausal symptoms—there is something here for every woman on the planet.

A third concept that I find compelling is the notion of using energy medicine to "evolve your body." Because our bodies evolved for a world that no longer exists, our adaptations are often imperfect compromises, and we can teach our bodies to adapt more effectively. This is a radical idea. Most people's sense of what is possible for them is tied to the prevailing notion called *genetic determinism:* "This is what runs in my family." "I am who I am because of my genes." But what we are discovering is that the same DNA can express itself in multiple ways. Your genes respond to the environment, and your body's energies are the first tier of that environment. Energy medicine shows you how to help those energies evolve, and when you do, your physical body will follow. An illuminating example from the book is that when Donna got her own spleen meridian into a better balance while addressing a serious health condition, she spontaneously lost 17 excess pounds without changing her diet. That you can evolve your body is a new and enormously exciting concept, and beyond merely offering the concept, energy medicine gives you a way to use it.

Energy Medicine for Women is extremely well documented. We now have the ability to measure the energy fields it discusses with scientific instruments. For the first time in human history, in fact, it is possible to scientifically establish that the Chinese physicians who intuitively mapped the acupuncture meridians thousands of years ago were right! Of course you don't need scientific instruments to use the techniques taught in this book or to feel their effects on your own body, but the scientific documentation builds a bridge between worlds that makes the book much easier for a medical school or a nurse or a doctor to use.

Hands-on energy medicine techniques are already starting to be used in conventional health care settings. Some hospitals routinely give patients an opportunity to have their energy fields balanced before and after surgery, chemotherapy, radiation, and other invasive procedures. It diminishes the shock to the body, hastens healing, and ameliorates many adverse effects, from pain to nausea to hair loss. In the medicine of the future, as I envision it, working with the patient's energy field will be the first intervention. Surgery will be a last resort. Drugs will be a last resort. They will still have their place, but shifting the energy patterns that caused the disease will be the first line of treatment. And before that, teaching people how to keep their energies in healthy patterns will be as much a part of physical hygiene as flossing or exercise.

It is significant that this book is directed toward women. The group with the most money and the most clout on the planet right now is baby-boom women in Western nations. And no group is more primed for changing established health practices. None of us wants to age the way we've seen our mothers age. And because we were the generation that said, "Don't trust anyone over 30," we don't take the previous generation's practices on faith. We also came out of the women's liberation movement, where we cultivated the idea that we could reinvent every stage of life we go through. We in fact have a knee-jerk reaction to the messages we've gotten about how one ages: "No, thank you. That's not about me. I won't do it that way." Energy medicine gives women tools they can use to reinvent themselves and what is possible for them, not only as they age, but also with their physical appearance, their vitality, and their physical health. And it gives new credence to the vision of "dying young as late as possible."

Being able to sense energy and work with energy is our native human territory. We are born with this ability, and people in indigenous cultures retain it. In technological cultures, however, where sensing the subtle energies of the body and the environment is not mentioned or cultivated, this ability has usually atrophied by the age of two. This needn't be the case. If infants are nurtured to see and feel the energy they are born knowing, they will retain this ability. And it is an enormously practical skill that will serve as a significant part of their inner guidance system. Fortunately, even if your natural ability to sense energy wasn't cultivated, you can learn how to work consciously with the body's energies, and this is very empowering. Donna, who literally sees subtle energies, has made a career of empowering people who don't see them to work with them effectively nonetheless.

And the implications are major. Not only can you improve your own health by using energy medicine to uplift the energies in your body, uplifting the energies in your body affects and uplifts the entire planet.

—JANUARY 2008
SANOVIV

Energy
Medicine
for Women

Introduction

What a piece of work is a woman!

—Mrs. Shakespeare

Your body's abilities to heal a wound, fight off disease, cope with stress, respond to danger, communicate to you what it needs, and reward you with pleasure when you provide what it requests all reflect an astounding intelligence that is totally independent of your mind. In this book, you will come to understand that this bodily wisdom is contained not primarily in the neurons of your brain but in energy fields that mobilize your cells into action, that coordinate the strategies used by your organs to maintain your health, and that bathe you in an atmosphere of peace and joy when things are well or in stress and alarm when they are not. You will also come to understand how this remarkable intelligence of your body and its energies is calibrated for a world that no longer exists, and that this is the underlying cause of some of our greatest health challenges. And most important, you will learn how to impact your body's energies and energy fields in ways that are attuned to the world in which you are living and that markedly enhance your health and well-being.

Energy is the living, vibrating ground of your being, and it is your body's natu-

1

ral self-healing elixir, its natural medicine. This medicine, this *energy* medicine, feeds body and soul, and attending to it restores your natural vitality. Energy medicine is the science and the art of optimizing your energies to help your body and mind function at their best.

The notion of taking action to enhance your health is familiar. You exercise, you are conscious that some foods and patterns of eating are better for you than others, you probably have considered vitamins or other dietary supplements, you take antibiotics when you have a persistent infection, and you may be on hormone replacement therapy if you are going through menopause or using other medications to help you with PMS or arthritis or anxiety or depression. This book will show you simple physical techniques, many requiring only a few invigorating minutes each day, that will rapidly shift the energies in your body. I present these techniques with confidence that they are among the most effective and efficient tools available for enhancing your health and boosting your vitality.

EMBRACING THE ARCHETYPAL FEMININE

We are living during one of the most precarious times in humanity's perilous history. A book about the energies that animate a woman's body is obliged to mainline you into a deeper connection with the archetypal feminine principle that must be re-embraced if our species is to survive. It needs to be more than just a guide to health. By personally embracing your deep feminine nature, you are more able to keep yourself upright and oriented through the chaotic cultural waters we are all navigating.

The archetypal feminine is not just a set of ideas and values somewhere out in the ethers. It is coded, yes, in your energies, but also in your genes, your hormones, and the actions they cause you to take. This book—by connecting you with the energetic foundation of the female body and soul—is designed to empower you to make better choices for your own health as well as to strengthen the connection between the precious wisdom of the archetypal feminine and your daily life.

Opening with "What a piece of work is a woman!" and its tongue-in-cheek attribution was not to slight men but to underline a cultural dilemma that has urgent consequences. We live within the context of a society whose patriarchal values are spinning out of control, marred by dominance over love, competition over cooperation, greed over justice, punishment over compassion, career over family, technology over nature, and war over peace. This tends to bring out the worst in both men and women.

At our core, however, our feminine instincts are toward love, cooperation, justice, compassion, family, nature, and peace. What a piece of work! If our deepest natures as women were to powerfully rise and flourish, the world would be propelled into adjustments it desperately needs. Learning how to bring your body into better health and balance through energy medicine allows your true nature to shine forth and is a training ground for becoming a force in transforming the energies around you so your world becomes a better place. And it is a path toward becoming an effective advocate for the love, cooperation, and peace that are our natural gifts to the world.

To best support you in this grand adventure, your body needs to be cared for and maintained, not only as a physical structure, but also as an energy system, which is exactly what it is!

AN EXQUISITE VIBRATION

As scientists are able to peer more and more deeply into the ingredients of physical matter, they have found that the unimaginably minuscule building blocks of nature, such as electrons and protons, are composed of ever smaller particles. Now scientists are speculating that at its base, matter may not be made of particles at all—it may be more like strings of vibrating energy.[1]

Whatever nature's most fundamental design, we do know that every electron, atom, molecule, cell, tissue, and organ vibrates, and its vibration determines much about its character and function.[2] Our entire bodies are, in fact, continually vibrating at a subtle level, moving energy and information through the connective tissues that encase us. We also resonate with energies in our environment. The frequency of the vibrations in your heart begins to match the frequency of the vibrations in the heart of your friend or lover or healer when you are in each other's presence. Vibration is everywhere, and appreciating vibration is at the core of energy medicine. At that core is also the realization that each of us is an exquisite vibration, an intelligent system of energies that supports the distinctive qualities that make us human: how noble in reason! How infinite in faculty! In action, how like an angel! In apprehension, how like a goddess![3]

GIVING EVOLUTION A BOOST

As remarkable as the intelligence of your body and the energies and energy fields that animate it, this intelligence, this guidance system, evolved for a planet that in many ways no longer exists. It is still attuned to an environment with no pollutants in the air or food, no crowded appointment books, no jet travel, no e-mail tyranny, no danger of not getting exercise, no conflict between your children and your career, no glass ceiling. Evolution could not possibly have anticipated the sedentary, information-saturated, polluted world in which your body is trying to thrive. Like a mighty knight from Arthur's court on a modern battlefield, many of your body's coping strategies are outmoded.

If you are a woman, the world your body evolved for kept you barefoot, pregnant, and probably dead before you reached menopause. Your hormones and neurotransmitters are the living record of your ancestors' adaptations to that world. While they worked well for your ancestors—you are the living proof—today they carry an antiquated game plan for directing your body's responses, sometimes in critical situations, such as illness or threat. If you want to thrive in the world you actually inhabit, a new set of adaptations, a new vibration beyond your body's ancient strategies, needs to be introduced. The underlying principle of this book is that by changing the vibration in your body, you can change the actions of its chemistry, helping it to better adapt to the strange new world in which we all live. You can *evolve your body* by changing its "energy habits."

Shifting an energy field can often trump preprogrammed biological strategies. Just like a radio signal creates the physical vibrations that result in the sounds that come through a speaker, changing an energy field exerts a domino effect on the body's chemistries. Simply thinking a happy thought or a sad thought instantly changes the chemicals being produced in the emotional centers of your brain. Energy medicine teaches you how to shift your body's chemistry as if you were handed a keyboard that produces electrochemical signals instead of sounds. Massage this energy point on your hand and you send endorphins to your brain. Tap that spot under your eye and your stomach cramps go away. Circle your hand over your chest and the heaviness in your heart lifts. Controlling your chemistry by managing your energies is the fast track for helping your body evolve and adapt to the challenges of the twenty-first century.

PROMISES

Energy Medicine for Women takes you on a personal tour of your body as an energy system, and it offers simple "energy tools" you can use to help your body function at its best. These tools are generally more precise, often more effective, and certainly less invasive than medication or surgery. They can also be self-applied, are available 24/7, and are free. Energy medicine shows you how to tap into your intuition, embrace your natural self-healing abilities, acquire new tools for assessing and optimizing your health, and generally become less reliant on pharmaceuticals or invasive procedures. While the marvels of modern chemistry can sometimes powerfully boost the healing process, energy medicine can still be used—even when medication is required—to ensure that you are using substances and dosages that are in harmony with your body's energies and needs.

Both creativity and wisdom are evident in your body's every action, whether it is to put on more weight than it seems to need, to take on a fever, or to rev up your anxieties for that big presentation. In recognizing what your body is attempting, you will be able to form a better partnership with its natural intelligence.

Energy medicine is particularly potent for women. For almost every health condition a woman faces, hormonal imbalances are in the foreground or in the background and, as you will see, energy medicine can help you dance with your hormones more effectively and gracefully than we in the Western world have been taught. This book will teach you how to become a better partner with your own energies and your hormones. When your energies are in harmony, your hormones follow. Energy medicine helps your body become more effective in adapting to the world in which you live, with its fast pace, stresses, pollution, and amazing opportunities to experience realms of human possibility that were not available to any generation before you.

BORN INTO A WORLD OF ENERGY

I see energy. I see it around people as clearly as you see the print on this page. I always have. It never occurred to me when I was growing up that this was unusual. My mother, sister, and brother all saw energy as well, and I assumed that everyone's take on life was based on the energies that they saw and felt. After all, how can you know if your friends need something they aren't aware of or are too ashamed to request if you can't see their energies? How do you figure out how to tap into your daughter's untapped potential with the necessary precision if you don't see it in her aura?

One of my earliest memories was watching from our front porch as my next-door neighbor, Sammy Henka, played in his yard. I was a tiny little thing. I could sit, and probably I could crawl, but I couldn't walk yet. Sammy was a couple of years older than I was; perhaps he was three or four. I saw Sammy coming off the porch, surrounded by a pretty azure-blue aura. He seemed carefree. Then I saw two girls coming down the street. A crisp amber color surrounded both of them, and there was something in the way their shared energy moved, pulsing in on them, that bothered me. I could tell by the thickness of the energy that they were talking about something with intensity. I also knew, even at that young age, that the energies I was seeing dealt with what I would later know to call "judgment." When the girls got to the house, they said in unison, "We don't like you, Sammy Henka. We don't like you!" Sammy's azure-blue energy instantly became brown, and it looked like it fell down an elevator shaft, descending to his knees, until he didn't even have an aura around his head or shoulders or torso. I died for him. They might as well have shot him through the heart. When I see such energy, I also feel it, and I couldn't conceive why human beings would do this to another person. Couldn't they see what they were doing? It was not until my early twenties, many years after this moment of painful empathy and shock, that I came to realize, *no,* they *couldn't* see what they were doing!

My ability to see and feel subtle energies has been my primary compass for relating to, understanding, and healing other people. The aura, as in the azure-blue energies that surrounded Sammy Henka, is but one level. The aura tells a lot about how the person feels, about the person's mental state, and even about the personality, but there is much more.

When I focus in with someone, I am often pulled straight into the energies that reside at the person's core by what feels like a gravitational force from within the aura. At that level, beneath the personality, beneath whether someone is loud or quiet or smiling or angry or kind or hateful, is pure goodness. We are each, at heart, an exquisite vibration! I can also see all the problems—mean, selfish, sloppy, senseless, or ugly defenses; harsh, brittle, jagged, stagnant, or mushy energies—but I don't stay focused there. I am drawn to the center, to the soul level. When you see into a person's essence, it is a magnificent moment. It takes my breath away. Every time!

Serving as the personal healer for so many individuals was a privilege that eventually became a problem. While I loved the healing process, as I matured, I found a growing urgency in myself to not perpetuate the false and disempowering idea that I was the one with the healing power and my client was the passive recipient. This concern, while valid on its own merits, was also driven by the simple fact that it had become impossible for me to see all the people who wanted sessions. I became determined to empower people to discover their own self-healing abilities, attune themselves to the energies that impact their health, and learn how to manage those energies for staying vital and healthy. After nearly a quarter-century with a full-time practice, at just before the turn of the millennium, I shifted my career in a way that has been exhilarating. I dedicated myself to teaching people how to help themselves and others using the powerful yet noninvasive techniques of energy medicine. My first book, *Energy Medicine,* is a by-product of that career shift. So are the almost 600 classes and talks I've given in the nine years between the publication of *Energy Medicine* and this writing. And so is this book.

I have had the privilege of witnessing again and again that it is possible to teach people who are not able to "see" energies to nonetheless work with their energies in ways that are powerful and often life-changing. *Energy Medicine for Women* presents basic methods that are helpful for *every* body and that also form the essential building blocks in tailoring a self-care program for your unique body, chemistry, and spirit.

ENERGY MEDICINE IN ACTION

An attorney who came to me for a treatment had been pushy in setting up the appointment. As the session began, she was abrupt, demanding, and skeptical. She was, in fact, exuding such aggression and hostility that I really couldn't see her essence. I invited her to lie down on the table, and I held some points to calm her energies. I could feel her start to relax a bit. Then waves of energy began to release from a vortex that sits over and reaches into the solar plexus. These energies were huge, overwhelming, and thick. I sensed that all her power issues were encrypted in these energies.

I began to work with this force field by moving my hand in slow circles over it. As the energetic storm began to clear, an opening appeared at the center of the vortex and I could see and feel into her energies more deeply. Because a person's history is literally coded in such energies, I began to see images. I saw her when she was in law school. She seemed like just a girl, young for her 22 years. She was being ostracized and ridiculed for being "soft." She learned that everyone is out for themselves and that you better find the other guy's Achilles heel before he finds yours. I laughed good-naturedly as I referred to her demeanor with me and told her, "I know it is your MO to nail the other guy before he nails you, but your energy shows that the reason for it is that you have really been hurt." She looked at me quizzically.

I described her experience in law school and named the conclusions she came to about being tough and invulnerable, and commiserated with her how it was perfectly understandable. She was listening attentively now, following my every word. I went deeper. More accurately, it was like the opening in her energy vortex pulled me in. It was a realm of such goodness and light, such a beautiful field, and so large, that I was overcome. I started to cry. She looked at me and said, "What's wrong?" I started to tell her what I saw. I described the glorious energy at her core. I depicted the images—embedded in her energy field—of a tender, innocent girl growing up in what felt to me like New Hampshire. I elaborated upon the dormant presence of the entirely appealing and likable person who was beneath her defenses. And I acknowledged the hard journey she had traveled. She felt my compassion as I explained why she had been a tormented soul for such a long time and why she had built all her defenses. Now the tears were in her eyes.

She lived out of town, but whenever she could travel my way, she would come in for a session. In essence, our work was to realign her with the deeper beautiful energy that was her true nature, her exquisite vibration. When I finally met her husband three years later, he gave me a long impassioned hug and credited me with having saved their marriage. She and I had never talked about their marriage.

Being pulled into direct contact with a person's divine nature is a profound experience. It transcends anything words can describe. The physical energies and colors fade into the background, and I know that I am in the presence of pure goodness. Suddenly I'm in a heavenly realm, enveloped by a radiance so deep and so beautiful that I know I am witnessing the face of God. I see profound wisdom in the person, and how truly good he or she is. I'm filled with love. This is my core orienting experience for understanding the world.

Whatever the personality, whatever the damage or defenses you encounter, you approach healing work differently when you are guided by a sense of certainty that unfathomed wisdom, goodness, strength, and beauty are the person's true nature. This is my orienting belief, and it is important to me to emphasize it from the start as we move into a narrower focus on energy flows, hormones, chemical imbalances, physical problems, and mechanical techniques. At the root of it all, in every one of us, is an exquisite vibration.

CREDENTIALS

What are my credentials to make all the claims and promises in this book? First, I am a woman. Women are rhythmically directed from within. A doctor with only one X chromosome cannot know what we know. We have an inside track on women's bodies.

Second, I am a woman who has had more health challenges than most, and I have overcome them. I was born with a heart murmur and serious metabolic problems. I had tuberculosis as a little girl, severe hypoglycemia and allergies into my 30s, multiple sclerosis by the time I was 16, a heart attack at 27, severe asthma in my thirties, a near-fatal insect bite at 33, and a malignant breast lump at 34. My entire body was breaking down, and I was told by more than one doctor to get my affairs in order. As it turned out, Western medicine giving up on me was one of the best things that could have happened. I was determined not to leave my young daughters for someone else to raise. Since Western medicine couldn't help me, I was forced to get very serious about helping myself. Today, in my midsixties, with the help of the techniques taught in this book, I am healthier and more capable of maintaining my health than I have been at any previous time in my life.

Third, I have struggled with my hormones since I was ten, including monthly cycles that could last two weeks and were debilitating. To complicate this, my body often had adverse or paradoxical reactions to standard medications and painkillers. I have had to figure out what would work for me in a culture that didn't give me answers or help.

Fourth, I have had a healing practice since 1977, applying my ability to see energies to helping more than 8,000 women (as well as several thousand men) in individual sessions.

Finally, I have successfully taught tens of thousands of people in energy medicine classes to work with their own health challenges. And, most important, I have figured out how to teach people who can't see energy to monitor their own energies and formulate interventions that have been effective in promoting their health and well-being.

Those are my credentials to offer this book. *Energy Medicine for Women* will show you how your body's energies affect every aspect of your life, and it will show you how to vitalize your cells, organs, and entire being by understanding your body as a network of energies that are responsive to simple physical techniques.

A "SHOW ME" ATTITUDE

I invite you to take a "show me" attitude throughout the book and think of its techniques as experiments. Often you will feel immediate results, even if they are subtle. Sometimes you will need to do an exercise on a daily basis for a while before you experience the desired outcome. And sometimes a technique isn't right for you. But because hands-on energy medicine is by nature noninvasive, no harm is likely to result by using the methods in this book as instructed. Side effects from energy interventions, though they rarely occur, may include unexpected emotional releases, mild nausea, headache, or other minor symptoms that quickly pass. The cause is usually too much energy being moved too quickly for a physically unstable body to readily accommodate it.

Though its methods are generally harmless and surprisingly potent, I am not suggesting that you should use this book as a substitute when professional health care is needed. Good self-help health books have the seemingly contradictory tasks of encouraging self-care *and* emphasizing that professional treatment is sometimes required. While there is an important place for each, the theme of this book is how much you can do for yourself.

Since the notorious Flexner report of 1910, when the medical community rose up to fight the threat of naturopathic and chiropractic physicians and began licensing the practice of conventional medicine, a line has been drawn in the sand for those who hold themselves out as health care professionals. There has been a virtual patent on the term "medicine." While at its best this line is meant to prevent quackery and charlatanism, it also has had the effect of concentrating the most prestigious and well-financed care into the hands of increasingly technological and pharmacological medical practices and away from folk *medicine* and health care practices adapted from other cultures, such as Native American *medicine* or traditional Chinese *medicine.* "Practicing medicine without a license" has come to mean any attempt at healing people that falls outside the framework of conventional Western medicine. It is illegal. And that is disgraceful!

I have personally been impacted by these laws. Early in my career, I was seeing many people who, as it turned out, were also patients at a group medical practice owned by five physicians who all had affiliations with the local hospital, which was adjacent to their office building. Enough stories got back to them that they decided to initiate legal proceedings against me for practicing medicine without a license (a felony). I can look back now and understand it from their point of view—this uncredentialed "healer" giving their patients, in their estimation, false hope by doing some kind of energy voodoo with passes of her hands over their bodies and other such nonsense. They subpoenaed several of these patients to testify, at a pretrial hearing, that I had been working with them on ways that crossed the line into their exclusive domain. Hearing from the witnesses, the judge was impressed that my work had indeed made a difference in their medical conditions. He asked the prosecuting attorney to find one patient who had been harmed. When he could not, the case was thrown out of court, with the judge agreeing that quackery should be prevented but then making some disparaging remarks about anyone trying to ban the kinds of healings the patients had described under oath. The incident had a redeeming sequel. Some eight years later, the hospital that was adjacent to the doctors' office had me teach an energy medicine class to its staff. Several of the doctors who had brought the charges against me attended and were among the class's most enthusiastic participants.

Our culture must come to grips with the fact that conventional Western medicine is failing us in many ways. According to one credible estimate, more deaths are caused by taking medications as prescribed and other iatrogenic reactions (illnesses induced by medical treatment) than all the other causes of death in the United States.[4] The health care system needs help, and time-honored, natural approaches that are powerful and safe are there to be embraced. While it places me and my colleagues in a legally more precarious position as we strive to stretch the boundaries of health care, we have chosen to reclaim the term *medicine* in calling our work "energy medicine." Your body's energies are the most natural medicine that exists, able to precisely orchestrate the body's self-healing capacities wherever required.

The conventional medical paradigm has a number of weaknesses. It is oriented toward diagnosing illness and treating disease. It is less oriented toward the larger concern of maintaining health. Energy medicine, at least as I and many others practice it, does not diagnose or treat illness. Rather, the focus is on assessing where the body's energies are blocked or not in harmony and then correcting the flow of these energies. Illness may provide clues about where the energies are disturbed, but it is not the focus of energy medicine. Energy medicine is every bit as much about optimal functioning as it is about overcoming maladies. This is a critical difference from the way we usually think about the word "medicine." In energy medicine, the "medicine" is not an external substance. It is the curative power of the proper flow of the body's energies. It is the most natural approach in the world. I think we should start training our children in its most basic procedures by the time they are able to wash their hands.

Where many interventions within conventional medicine are "done to you," you can do many energy medicine procedures yourself. The following pages are filled with techniques that have evolved from three decades of working with thousands of women on the issues being addressed. However, one of the most serious limitations of energy medicine is that if you don't use it, it doesn't work. Yes, sad but true. If you just read about the techniques and don't do them, the book will still be informative, and there is value in that, but it will be a little like hoping you will lose weight by putting the diet book under your pillow instead of doing the diet.

HOW TO USE THIS BOOK

A dilemma in writing this book has been that it is intended for two audiences: people who have read my first book, *Energy Medicine* (or who are otherwise familiar with the material it presents), and people who are new or relatively new to energy medicine and my approach to it. In order to apply energy medicine for women, the reader first needs some background in energy medicine. So I have incorporated essential concepts and methods from my first book into this book and ask for indulgence from readers already familiar with this material. I have also tried to present these basics in a manner that will be a useful review for those already conversant with them (this material is found primarily in chapters 1 and 2 and the Appendix). I should also mention that the first book gives you more tools to figure out ways to apply energy medicine to your own unique body and health concerns, while this book, by necessity, is more formulistic, offering techniques that focus on many of the generic health issues women frequently face.

You can approach *Energy Medicine for Women* in several ways. If you are new to energy medicine, my suggestion is that you first read chapters 1 through 3 and immediately begin to experiment with the procedures they present. You will gain a good understanding of how the body's energies govern a woman's health, chemistry, and spirit, and how simple exercises can optimize that process. Next, read the text of subsequent chapters that interest you and scan the procedures. Use those that apply to you, and then keep the book handy as a self-help reference.

Chapters 1 and 2 provide an introduction to energy medicine and its most basic methods, synthesizing some of the most important principles and practices from my earlier book. If you have studied the first book, you may certainly skip these chapters, and the Appendix, though I think you will still find they provide a good review and synthesis oriented toward the theme of this book. Chapter 3 gives you a more detailed understanding of the relationship between your hormones and your body's energy systems, as well as the myths and the medical politics that affect a woman's health, and it teaches several basic energy medicine procedures for working with the hormones involved with stress and with immune function. The subsequent chapters then present an energy medicine approach to women's issues such as menstruation, sexuality, fertility, pregnancy, birth, menopause, and weight management, offering techniques you can apply directly for each area of concern. Finally, the Appendix is a tutorial in the important art of energy testing, which can help you tailor the entire program to your body's unique biochemistry and needs.

The physical techniques presented throughout this book are each designed to bring about specific benefits. I have tried very hard to describe each procedure clearly, and they are well illustrated with Christine Alicino's superb photographs.

The techniques each require only from a few seconds to a few minutes, and I've generally indicated how much time will be needed. Please do not be put off by the length of the written instructions. Spelling out the procedures in a precise, step-by-step manner often requires instructions that *take longer to read than to do*. But by doing them as you read, you will find that you can pick up each technique quite readily. I have also made a companion DVD program (available through www .EnergyMedicineForWomen.com) that leads you through most of the techniques presented in this book. It simply is the case that it is easier to follow along on your television screen than to follow written instructions. If you go to that site, you can also watch a free video of me introducing you to this book. I know that if I can first hear and see an author, I have a more visceral relationship as I read the book, so do let me personally introduce myself using the magic of Internet video.

As with my first book, *Energy Medicine,* this book also lays out a systematic yet open approach for consciously working with your own energies, or the energies of those you care about, in order to achieve a healthier body, sharper mind, and more joyful spirit. As was my approach in that book, I have generally not pulled my punches in presenting techniques that are potent and that can be helpful with difficult conditions. On the basis of my experience in teaching these techniques to more than 50,000 people, I have a good idea of how each method presented here impacts a wide range of people, and I have used my best discretion. Your part of the bargain is to exercise your own best discretion as you go through the program. If you need professional treatment, please consult a competent health practitioner! The techniques in this book will only complement the care you receive. But while sincerely stressing the value of professional intervention, I do not want to undercut my primary message and deep conviction that there is much you can do to care for yourself. Ultimately, it is you who is responsible for your health, and the more you know and the more you do for yourself, the easier the task for those who care for you professionally, and the better your health will fare.

In the acknowledgments to *Energy Medicine,* I noted that my husband, David Feinstein, "has been tireless in interviewing me and writing first drafts from those interviews, editing transcripts of tapes from my classes, conducting computer searches, and generally bringing left-brained organization to my right-brained way of being. His abilities to craft a phrase, suggest an analogy, find order amid complexity, and place an idea into its larger intellectual context, all the while retaining the spirit of *my* voice, permeate this volume. In short, this is the book *I* would have written if my mind worked like both of our minds combined." While David has also been a major force in the creation of this book, I am pleased to say that a decade later, each of our minds works a bit more like our combined minds used to work. Some of the tools we offer here have served us both in that regard. Balancing your energies balances your mind. May this book help you find your way to greater balance, health, and inner peace.

Chapter 1

A Medicine Called Energy

We're in the midst of a radical, essential paradigm shift in the Western
medical model. . . . It is imperative that we expand our concepts of
health care to include subtle energy and energy medicine.

—JUDITH ORLOFF, M.D.
Positive Energy

Energy medicine is probably the "real" oldest profession. Knowing how to keep the body's energies healthy and vital gave our ancestors a tremendous edge for surviving in the wild. Strategies for sensing and correcting energy imbalances by stimulating specific "energy points" have been passed down through generations in China, as well as in other parts of the world, for at least 5,000 years. A body that had been mummified in a snowbound mountainous region along the border between Austria and Italy around 3000 B.C. had tattoos on exactly the points that are indicated in traditional acupuncture for treating the kind of lumbar spine arthritis revealed by an X-ray analysis of the body. Nine of the fifteen markings were along a meridian that is used in treating back pain, including one on the precise acupuncture point that is considered the "master point" for back pain. Forensic analysis also revealed that the body's intestines had been rife with whipworm eggs, and indeed, some of the other markings were on points that are traditionally used for treating stomach upset.[1] Similar tattoos have been found on mummified bodies in other regions, ranging from South America to Siberia.

In traditional Chinese medicine, acupuncture and other methods were used to keep the body healthy by keeping the energy fields that support it healthy. In some provinces of ancient China, you paid the physician a nominal fee when you were well. If you got sick, the physician would work hard to try to cure your illness, but you did not have to pay because the physician had failed to keep your energy field healthy enough to prevent the illness. Try proposing that arrangement to your local hospital!

Disturbed energies precede illnesses in the physical body, and it is possible to correct energy imbalances *before* they coagulate into illness. Energy medicine can address illness, but it can also prevent it. Think of the energy field as the blueprint of the physical body. If this living blueprint remains sound, the body stays healthy. If the blueprint is damaged, the body follows. Maintaining a healthy energy field is a powerful strategy for maintaining health and preventing illness. By adeptly engaging the energy systems within your body, you can improve your health, turn around an illness, and live with greater vitality day by day.

TWO SIMPLE ENERGY EXPERIMENTS

The tools of energy medicine range from elegantly simple to highly sophisticated. Some of the methods involve touching the body, while others involve no touch at all. Try this experiment. Bring your palms toward each other until they are about three inches apart. Now twist your arms so they form an X, with your wrists at the center of the X, still about three inches apart. Bring your attention to the space between your wrists. Because your wrists contain several energy centers, the energies will connect, and most people will feel some sensation in the area between their wrists. Move your wrists about an inch closer and then out a few inches back and forth. What happens to the sensations in the space between your wrists as you change the distance? Do not be concerned if you do not feel the energy, but it is there. The palms of your hands also emit a considerable amount of subtle energy. See if you can feel the energy in the space between your hands if you cup your hands and move them toward and away from each other.

Now do another experiment. Most people carry some tension in their shoulders. Reach your right hand over to your left shoulder and press into any point with your middle finger. Feel around until you find the spot that is most tender (if there is another area of your body you would prefer to focus upon, do so). Give a rating of 0 to 10 to this spot, with 0 meaning no tension or tenderness at all and 10 meaning extreme tension or tenderness. Next, vigorously rub your hands together and shake them off. Everyone's hands have a measurable energy field. Now cup your right hand about two inches above the area where you felt tension and begin to move it over the area in a slow counterclockwise circle, making about a dozen rotations. Notice if you have a sense of energies being exchanged between your hand and your shoulder. Hold your hand over the area a few more seconds, then relax. What sensations do you feel? Now press the original point with your middle finger and again rate it from 0 to 10. Most people find that the tension diminishes.

While about as simple as a technique could be, this experiment shows how readily we can move our energies. Energy medicine begins with such simple tools for reducing tension and fostering healing, and it progresses to more complex protocols for bringing about targeted benefits and addressing serious health challenges.

YOUR HEALING IS IN YOUR HANDS

Energy medicine is one of five domains of "complementary and alternative medicine" identified by the National Institutes of Health (NIH).[2] Methods used within energy medicine include electrical devices, magnets, crystals, needles, aromas, and herbal or other ingested substances. But the tool used by the largest number of practitioners for moving and harmonizing the body's energies and fields is the human hand. Practitioners as well as anyone reading this book can use their hands to bring balance and harmony to the body's energy fields. You can tap, massage, pinch, twist, or connect specific energy points on the skin. Because everyone's hands carry a measurable electromagnetic charge, specific areas of the body can be surrounded with the hands to produce a field effect, or the hands can be used to move and align the body's energies by tracing specific energy pathways along the skin. The use of postures and movements are other noninvasive ways to benefit the body's energy system.[3]

One of the most fundamental (if not always honored) axioms of conventional medicine is that the least invasive measure likely to impact an illness should be the first applied. Fortunately, low-tech energy medicine procedures are not only readily available, they are also noninvasive, preventive, and strikingly cost-effective in contrast to the skyrocketing expense of conventional medicine and its disastrous impact on the economy.

The power of energy medicine over conventional medicine's "medicate or cut" approach is also seen in the way its methods are able to rapidly affect your entire body, and your energy "blueprint," rather than focus only on body parts. How are energy interventions able to have this "holistic" influence?

While we don't usually think about the body's connective tissue as an organ, it is actually a remarkable one, and it transmits energy impulses to reach every part of the body virtually instantly. Each of our organs, according to Dawson Church, *"is encased within the body's largest organ, which functions as a liquid crystal semi-conductor"* that conducts information as well as electrical signals by being able "to *store* energy, *amplify* signals, *filter* information, and *move* information"[4] to every cell of your body. With the connective tissue acting as a giant electrical semiconductor, energy interventions can simultaneously be brought to every cell of the body.

Six fundamental principles that are the basis of energy medicine give it strengths not found in conventional health care models. Further elaborated in an article that is available as a free download (www.EnergyMedicinePrinciples.com), these principles include:

1. **Reach:** Energy medicine addresses biological processes at their energetic foundation so that it is able to impact the full spectrum of physical conditions.
2. **Efficiency:** Energy medicine regulates biological processes with precision, speed, and flexibility.
3. **Practicality:** Energy medicine fosters health with interventions that can be readily, economically, and noninvasively applied.
4. **Patient empowerment:** Energy medicine includes methods that can be utilized on an at-home, self-help basis, allowing more effective self-care while also fostering a more creative patient-and-practitioner partnership in the healing process.
5. **Quantum compatibility:** Energy medicine adopts nonlinear concepts consistent with distant healing, the healing impact of prayer, and the role of intention in healing.

6. *Holistic orientation:* Energy medicine strengthens the integration of body, mind, and spirit, leading not only to a focus on healing, but also to achieving greater well-being, peace, and passion for life.

SENSING SUBTLE ENERGIES

In short, it is time for Western medicine to embrace the energy paradigm and move forward as a more powerful, attuned, and responsive discipline.

As you begin to apply energy techniques in your life, you open yourself to a realm of reality that operates beneath the radar for most people in technological cultures. The subtle energies in the environment, however, were an essential source of information for our ancestors, telling them if danger lay around the next corner or if a particular plant could be safely ingested. The subtle energies within their own bodies guided their daily choices and activities as decisively as the not-so-subtle energies of a headache might cause you to take a nap.

I am certain that when babies are born they are far more able to register these subtle energies than we are as adults. Have you ever noticed the way infants will often look intently just above your head or to the side of your face? There is no question in my mind that they are seeing the energies that surround you. Babies see energy, feel it, sense it, know it. But because the brain has so much to learn, and because the realm of subtle energies is rarely spoken of and rarely validated, these sensitivities become dormant. They fall out of the loop in the learning process.

I have on occasion, however, had an opportunity to encourage a pregnant woman or the parents of a baby to talk with their child about energy from the beginning. Regardless of whether or not the parents see energy, I ask them to imagine the energies that animate all of life and talk about them. The parents may be speaking only about their imaginings, but they are attuning themselves to what is a reality for their child. From the offspring of these parents, I now know a handful of older children who can still see energy in vivid colors and are able to talk about it freely and easily. After having been mostly on the road teaching for seven years, I had returned to the town where I had had my private practice. I was walking on the main street and, in the opposite direction, five large young men who looked like high school football players were sauntering in my direction. I did not recognize them and they looked a bit intimidating. When we were about half a block apart, one of them, who as it turned out had been my client along with his family when he was a child, looked at me directly and called out for everyone, including his buddies, to hear: "Hey, Donna!!! You still have pink in your aura!" Welcome to Ashland, Oregon, dear reader. I have also witnessed many adults as they began to see and accurately sense energies after experimenting with techniques presented in this book. They are opening to a deeply ingrained but forgotten ability.

Again in Ashland, a woman dragged her physician husband to one of my weekly evening classes. He found the idea of energy healing implausible and the concept that a person could see color in people's energies ludicrous. But as he delivered his sometimes sarcastic comments about "seeing" colors, he spoke in one of the deepest and most beautiful voices I had ever heard. When people paired up to practice techniques with each other, he would always work with his wife, back in a corner of the room. On the last evening of the class, as people were practicing with partners, a high-pitched, squeaky voice from the back of the class exclaimed, "Purple, I see purple!" It was the physician. He was so surprised by seeing the purple in his wife's energies that his voice jumped three octaves. This jolt to his worldview resulted in his taking further classes, now on his own motivation, and he began to incorporate energy medicine into his medical practice.

When you learn the native language of your own unique energies and energy fields, you become able to read them, hear them, and converse with them. One of the trickiest things in cultivating new sensibilities about your own and others' energies is that they often don't show up in the ways you expect them. Like synesthesia, where certain people can *smell* colors or *see* sounds, the perception of energy may just slide into one of your normal sensory channels. I've known people who can hear, smell, or taste specific energies, rather than see or feel them as I do. My own sense of taste has actually also grown stronger. I can usually taste which of the five elements (the elements are one of the core energy systems) is a dominant theme in the health challenges, as well as psychological issues, a person is moving through. For instance, a metallic taste in my mouth alerts me to an imbalance in the metal element. I've now known several people who can *hear* energies move and detect where they are blocked in a client lying on their treatment table. One of my colleagues began to smell energies, and the scents became so overwhelming that she had to get out of the work for a time. You cannot know in advance how you will register subtle energies. We all have different strengths and different ways of knowing. As you expand into the language of energy, the only thing you can predict for sure is that it will speak to you in the way it chooses, not necessarily in the way you hope for or expect.

My friend and associate Sandy Wand often sees symbols when she is working on someone. She never knows where they will take her. But she has learned to describe them to her clients, often having no idea what they mean but trusting that they will eventually begin to make sense. I had not told her of a frightening experience I had had in which I thought I was dying. Lying in my hotel room bed in London, I felt that all my energy instantly dropped, like a runaway elevator, down into my root chakra, stopping with a jerk. Suddenly, I couldn't see the room anymore. All I could see was a deep blue-black that felt like hot ink boiling within my root chakra. It began to rise, filling my entire body. It felt like a poisonous fluid was inside me.

When I returned home, still feeling poisoned, Sandy gave me a session. After working on me for a few minutes, she said, "Well, this isn't going to make a lot of sense, but you know how squid squirt out ink to protect themselves? I'm getting an image that, like a squid, your root chakra has been squirting out energy in order to protect you." The piece of the puzzle she gave me that I didn't have was that while I thought this deep blue-black was an energy of death, she saw it as quite the opposite. This was an energy of life protecting itself. Squid squirt out ink so nothing can see them or get to them. This provided me with an extremely useful insight. If I didn't start protecting myself better, *then* I might be dead. Since I wasn't succeeding all that well at setting boundaries, my energy system was trying to set a boundary for me. The blue-black "ink" was putting out a force to both contain my own energies and keep out the energies of others who might harm or drain me. Sandy has a gift, but when she first started doing energy healing, she didn't know she had it. Open to the unique ways energies reveal themselves to you, and your own natural capacities to work with energy can flourish as well.

BETTER LIVING THROUGH ENERGY

A happy bit of fallout since the publication of my first book has been how many times someone I've never met or even seen in a class has sent our office an e-mail describing how using the methods in the book have turned around a difficult physical condition. Just today, September 21, 2007, as I was working on this section, two such e-mails were forwarded to me (and I have subsequently received permission to include them here). In the first, a woman with cystic fibrosis who had received a double lung transplant wrote to the person who ships our books and DVDs: "There seems to be a big buzz on the CysticFibrosis.com site about *Energy Medicine*. I thought you all would appreciate what others are saying." Among the comments in the interactive discussion blog we found when we went there:

I've mentioned before the book Energy Medicine. *WHY do I love it so much? For so many reasons. I learned how to energy test in ways that I did not know before. I know how to test for foods, vitamins, and drugs that work for my body through energy testing. That is in the book. My latest discovery. I would consider myself borderline diabetic. I took my glucose after eating one day and it was 143 (one hour after eating a high-energy bar). I practiced the daily routine and traced all my meridians and after eight minutes, I retested my glucose again to see a change. My glucose dropped nineteen points. After eight minutes that is pretty remarkable. I did this for three days in a row to see if this was for real, and every single time I had the same results. Two days ago when I went for my six-month checkup, I told three doctors about the blood sugar readings and they all seemed very impressed. Before I eat now, I do all of the energy routines that I mentioned. I have had the book now for four years, and I have read it over and over again.*

The very next person to post an entry, another double-lung transplant recipient, described how she has also

tried the Energy Medicine *book and some exercises to help with things like headaches, cramps, constipation, and gas pains. For me the book and its exercises are amazing. I can tell a significant change in my heart rate or blood pressure when I do and do not do the exercises. I have actually checked my blood pressure when it was high—like in the 150/90 range—and then did a few of the energy medicine techniques and checked it again within about five minutes and it had dropped to about 120/80, into a normal range.*

The second e-mail to arrive on September 21 stated:

I am writing this to express my sincere thanks to Donna and David for their generosity in sharing the invaluable knowledge of energy medicine and energy psychology. I was recently diagnosed with dengue—an epidemic spread by mosquitoes, which has no vaccines or drugs to treat it, squeezes the life out of the victim through its symptoms, and is fatal too. I was admitted into the hospital. My blood tests showed infection in blood and liver. My doctor was very concerned as the liver was affected. Hearing this, all family members got extremely worried. I was just the opposite. I asssured them that I would be out of the hospital in two to three days (something considered impossible for someone with my condition). I do not know why I said that, it just popped out of my mouth. But that's exactly what happened! How?

Day 1: I recalled what Donna had said in her DVDs: "Disease never lives in a well-oxygenated body." "Strengthen your spleen to be healthy." I had only one hand free as the other was on tubes for intravenous fluid, so the challenge was how do I do all these exercises with one hand? I could breathe, buzz my K-27, thump my thymus, and tap my spleen points on one side and then the other with a single hand, but couldn't do the rest. I remembered Donna mentioned a man who had a stroke and got well by just imagining doing the routine. I did the same. I visualized myself flushing spleen, hooking up, "connecting heaven and earth," drying off, and doing figure eights all over where my single hand could not reach. I also did energy psychology. My "karate chop" was with my single hand on my thigh! I focused on more figure eights on my stomach, as I had severe vomiting. I also sat there appreciating every single part of my body for being healthy and all the objects in my room (including my vomiting pan) for being there and all the people behind them. I remembered Donna's words: "Gratitude is the best vaccination for everything" [gratitude is not just something you do with your mind—it activates a healing energy called the radiant circuits, discussed later in this chapter].

My vomiting disappeared within 12 hours. Everyone was surprised to see me eating my favorite food. My fever was under control, and I felt so energetic that I didn't sleep at all except the few hours that I got at night between routine blood samples and temperature checks every two hours. My doctor was so surprised that he said, "You look so bright! What have you been up to?"

Day 2: *Blood results in the morning showed higher platelet count, almost normal. I kept doing all that I did on Day 1. During his evening visit, my doctor told me that my blood results and temperature were almost normal. He was surprised to learn what I ate and how energetic I was.*

Day 3: Completely *recovered and discharged from the hospital!*

No one could believe it. Patients with dengue are usually bedridden and take an average of seven to ten days to fully recover, if they recover at all. The doctor was so curious that he asked, "Do you do any kind of exercise or something?" I immediately shared with him about energy medicine and how I learned it. As I left, he said with a smile, "I'll remember, energy medicine!" My mother, who was a nonbeliever, is now doing the daily five-minute routine [see page 51]. I am yet to give her the gift I bought her, Donna's Energy Medicine Kit. My husband and daughter are staunch believers of energy medicine. My brother is planning to make a special visit just to learn this from me. I am so happy and thankful!

While single cases may simply be "spontaneous remissions" having nothing to do with energy medicine, you can imagine how gratifying it is to consistently hear comments like these on a full spectrum of health concerns from people who have no particular investment in "proving" that energy medicine works, and you may find it encouraging to know that such cases are not unusual as you begin to apply the methods presented in this book to your own health concerns.

Energy medicine works with the life force, and part of its power is this focus on the well-being of the entire body rather than merely on parts or symptoms. Western-trained physicians, on the other hand, learn anatomy on cadavers—the flesh without the force, the limbs without the life. Because conventional medicine focuses on physical structures rather than energy, its primary tools are those that work on matter: drugs and surgery. Because energy medicine focuses on the *forces that animate* the physical body, its tools are more subtle. They are generally noninvasive. And they are often more effective. Energy is also free, user-friendly, and always available to help us feel better and function better.

But if that is so, why isn't energy medicine mainstream? Why is this book shelved in the Alternative Medicine section? Conventional medicine is a trillion-dollar (that is one thousand billion) industry annually. The total outlay for alternative medicine is two-tenths of one percent of that.[5] In business, money talks louder than other truths and, make no mistake, the health care industry is *big* business. Energy medicine, being quick, effective, and cheap, can threaten that business. Extraordinary claim? Yes!

And I am not denying the powers and wonders of modern medicine. Nor am I claiming there is a conspiracy to keep you from knowing about energy medicine. I am claiming, rather, that there are many, many areas where conventional medicine is not particularly proficient in helping you maintain your optimum health and that there are market forces and cultural forces that have kept energy medicine—which may be far more effective and safer in many of these areas—from becoming mainstream.

But a new paradigm is rapidly emerging in which the body's energies are going to become a central focus in the practice of medicine. According to Mehmet Oz, M.D., one of the most respected surgeons in the United States and the director of the Cardiovascular Institute at the Columbia University College of Physicians & Surgeons (speaking to an international audience on *Oprah*, no less): "The next big frontier in medicine is energy medicine." Dr. Oz is not alone in this opinion. Norm Shealy, M.D., founding president of the American Holistic Medical Association, has flatly stated that "energy medicine is the future of all medicine." Richard Gerber, M.D., predicts that "the ultimate approach to healing will be to remove the abnormalities at the subtle-energy level which led to the manifestation of illness in the first place."[6]

This emerging paradigm is actually both old and new. According to Albert Szent-Györgyi, Nobel Laureate in Medicine: "In every culture and in every medical tradition before ours, healing was accomplished by moving energy." I sketch some of the basic elements of this emerging paradigm in the remainder of this chapter. However, if you are already familiar with it, are not interested, or are eager to get on with the book's self-care program, you may proceed directly to Chapter 2. But I think you will find it quite interesting to learn about the science that is supporting the idea that health is influenced in profound ways by the invisible energy anatomy of the human body.

THE BODY'S ENERGIES

While energy takes many forms—such as kinetic, thermal, chemical, and nuclear—the energies most pertinent to energy medicine seem to involve the body's *electrical* energies, *electromagnetic* energies, and *"subtle"* energies:

- Like a miniature battery, each cell in your body stores and emits *electricity.* Every breath you take, every muscle you move, and every morsel of food you digest involves electrical activity.
- Wherever electricity moves, *electromagnetic fields* are produced, and I explore the role of such fields in health and healing in the following discussion.
- *Subtle energies* were described by Einstein as energies we know of because of their effects but do not have the instruments to detect directly. While these subtle energies cannot move a needle on a gauge, many healers know how to engage them to restore health and vitality. Interestingly, a device developed at Stanford University that detects a form of energy that until recently had eluded scientific instruments shows that this energy responds to human intention.[7]

Other energies are also at work in the body, though they are not so much the focus of energy medicine. The forces that hold an atom's nucleus in place, for instance, are more than ten billion billion billion times stronger than gravity. If you start to feel droopy in the afternoon, it may be inspiring (though counterintuitive in that moment) to recall how much energy sits in every one of your cells.

I see nine energy systems in the human body, though I was not the first to discover any of them. Each is recognized in the healing tradition of one culture or another and each is a mixture of electrical, electromagnetic, and more subtle energies. For instance, the chakras, one of these nine systems, can be measured according to electromagnetic frequencies in the area of the body where the chakras are located.[8] But the chakras are not *just* electromagnetic frequencies. They also contain information that a sensitive person can "read" intuitively by becoming attuned to the subtle energies held by that chakra. That is why a healer may see or even relive a person's deep traumatic memories by tuning in to the person's chakra energies.

Not only can these visions be immediately verified by the client, different healers working with the same person often pick up on the same story. Is a chakra an electromagnetic field? Yes. Is it also a more subtle type of energy that carries information electromagnetic fields are not known to contain? Yes again.

While I have always seen these nine energy systems, it was only from working with people in my practice that I clearly distinguished among them and learned that each has been identified, named, and worked with in the healing traditions of other cultures. Of the nine energy systems, names for three of them have entered our language: the meridians, chakras, and aura. Descriptions of these energies by "seers" often correspond with one another,[9] and their physical existence is increasingly being verified by instruments measuring electricity, electromagnetic fields, light, or other forms of energy. The following overview of the nine systems is built around an *analogy* for each system, designed to give you a more concrete sense of the nature and function of these invisible systems.

1. *The meridians:* In the way an artery transports blood, a meridian transports energy. As the body's *energy bloodstream*, the meridian system carries the life force, adjusts metabolism, removes blockages, and even determines the speed and form of cellular change. The flow of the meridian energy pathways is as critical as the flow of blood. No energy, no life. Meridians affect every organ and every physiological system, including the immune, nervous, endocrine, circulatory, respiratory, digestive, skeletal, muscular, and lymphatic systems. Each system is fed by at least one meridian. If a meridian's energy is obstructed or unregulated, the system it feeds is jeopardized. The meridians include fourteen tangible channels that carry energy into, through, and out of your body. Your meridian pathways also connect hundreds of tiny, electromagnetically distinct points along the surface of the skin. These are known as acupuncture points. They have less electrical resistance than other areas of the skin, and they can be stimulated with needles or physical pressure to release or redistribute energy along the meridian pathway.

2. *The chakras:* The word chakra translates from the Sanskrit as "disk," "vortex," or "wheel." The chakras are concentrated centers of energy. Each major chakra in the human body is a center of swirling energy positioned at one of seven points, from the base of your spine to the top of your head.

Whereas the meridians deliver their energy *to* the organs, the chakras bathe the organs *in* their energies. Each chakra supplies energy to specific organs, corresponds to a distinct aspect of your personality, and resonates (respectively, from the bottom to the top chakra) with one of seven universal principles having to do with survival, creativity, identity, love, expression, comprehension, or transcendence. Your chakra energies encrypt your experiences in much the way your neurons code your memories. An imprint of every emotionally significant event you have experienced is recorded in your chakra energies. A sensitive practitioner's hand held over a chakra may resonate with pain in a related organ, congestion in a lymph node, subtle abnormalities in heat or pulsing, and areas of emotional turmoil, or may even tune in to a stored memory that might be addressed as part of the healing process.

3. *The aura:* Your aura (or biofield, the term used by scientists who have been studying it[10]) is a multilayered shell of energy that emanates from your body and interacts with the energies of your environment. It is itself a *protective atmosphere* that surrounds you, filtering out many of the energies you encounter and drawing in others that you need. Like a space suit, your aura protects you from harmful energies. Like a radio antenna, on the other hand, it brings in energies with which it resonates. The aura is a conduit, a two-way antenna that *brings in* energy from the environment to your chakras and that *sends* energy from your chakras outward. When you feel happy, attractive, and spirited, your aura may fill an entire room. When you are sad, despondent, and somber, your aura crashes in on you, forming an energetic shell that isolates you from the world. Some people's auras seem to reach out and embrace you. You know such people, and if you could see their auras, you would find that their aura energies do indeed reach toward your own. You also know people whose auras are so tightly bound and protective that they keep you out like an electric fence. A study conducted by Valerie Hunt, a neurophysiologist at UCLA's Energy Fields Laboratory, correlated "aura readings" with physiological measures.[11] The auras seen by eight practitioners not only corresponded with one another, they also correlated with wave patterns picked up by electrodes on the skin at the spot that was being observed.

4. ***The electrics:*** The electrics are an energy that seems to emerge from the electrical dimension of the other energy systems. They are not an independent energy system like the meridians, chakras, or aura, but they are rather intimately related to all the major energy systems: separate from each but also an aspect of each, something like the way liquid is separate from yet part of each of your organs. The electrics serve as a bridge that connects all the energy systems at the basic level of the body's electricity. I usually have no idea what will occur when I first hold a person's electric points. The energy goes where it is needed. There are reports of scar tissue being healed during an electrics session, heart ataxia eliminated, a knee replacement operation avoided, and all manner of emotional trauma overcome. But most important, in terms of whole-body healing, is the way the electrics connect all the systems. If energy fields such as the aura and chakras align the organs and other energies by surrounding them, the electrics move right through them, connecting and coordinating them at the tangible dimension of their electrical nature.

5. ***The Celtic weave:*** The body's energies spin, spiral, curve, twist, crisscross, and weave themselves into patterns of magnificent beauty. The equilibrium of this kaleidoscope of colors and shapes is maintained by an energy system known by different names to energy healers throughout the world. In the East, it has been called the "Tibetan energy ring." In yoga tradition, it is represented by two curved lines that cross seven times, symbolically encasing the seven chakras. In the West, it is seen in the caduceus, the intertwined serpents on a staff—also crossing seven times—associated initially with the Greek god Hermes, messenger for the gods, and later used as a symbol in alchemy and then medicine. I use the term *Celtic weave* not only because I have a personal affinity with Celtic healing, but also because the pattern *looks* to me like the old Celtic drawings of a dynamic, spiraling infinity sign, never beginning and never ending and sometimes forming a triple spiral. Like *invisible threads* that keep all the energy systems functioning as a single unit, the Celtic weave networks throughout and around the body in spiraling figure-eight patterns. It is a living system, continually weaving new crossovers, ever expanding and contracting. The double helix of DNA is this pattern in microcosm. The left hemisphere's

control of the right side of the body and the right hemisphere's control of the left side is this pattern writ large. These crisscrossing energies permeating your body are the "connective tissue" of your energy system.

6. *The five rhythms:* Your meridians, chakras, aura, and other essential energies are influenced by a more pervasive energy system. I do not see it as a separate energy but rather as a *rhythm* that runs through all the others, leaving its vibratory imprint on physical attributes, health patterns, and personality traits. Mapped long ago in traditional Chinese medicine, all of life was categorized into five "elements," "movements," or "seasons" (there is no perfect translation—all three terms have been used, suggesting qualities of being substantial, dynamic, and cyclical). These energies were considered the building blocks of the universe, providing a basis for understanding how the world works, how societies organize themselves, and what the human body needs to maintain health. Metaphors for describing these five distinct rhythms have drawn from concrete, observable elements of nature (water, wood, fire, earth, and metal) and from the seasons (winter, spring, summer, Indian summer, and autumn). Like the background music during a movie, the person's primary rhythm, in combination with the changing rhythms of life's seasons, directs the tone and mood of the entire energy system and sets the atmosphere of the life being lived.

7. *The triple warmer:* Triple warmer is the meridian that networks the energies of the immune system to attack an invader, and it mobilizes the body's energies in an emergency for the fight, flight, or freeze response. It operates in ways that are so beyond the range of any other meridian that it must be considered a system unto itself. Its energies work in conjunction with the hypothalamus gland, which is the body's thermostat and also the instigator of the body's emergency response. Like an *army,* triple warmer mobilizes during threat or perceived threat, coordinating all the other energy systems to activate the immune response, govern the fight/flight/freeze mechanism, and establish and maintain habitual responses to threat.

8. *The radiant circuits:* Whereas the meridians are tied to fixed pathways and specific organs, the radiant energies operate as fluid fields, embodying a distinct spontaneous intelligence. Like hyperlinks on a Web site, they jump instantly to wherever they are needed, bringing revitalization, joy, and spiritual connection. If triple warmer mobilizes your *inner militia,* the

radiant circuits mobilize your *inner mom,* showering you with healing energy, providing life-sustaining resources, and lifting your morale. The radiant circuits function to ensure that all the other energy systems are working for the common good. They redistribute energies to where they are most needed, responding to any health challenge the body might encounter. In terms of evolution, the radiant circuits have been around longer than the meridians. Primitive organisms such as insects move their energies via the radiant circuits rather than through a meridian system, and the radiant circuits can be seen in the embryo before the meridians develop. As in the way that riverbeds are formed, it is as if radiant energies that habitually followed the same course became meridians.

9. *The basic grid:* The basic grid is your body's foundational energy. Like the chassis of a car, all the other energy systems ride on the energy of the basic grid. For instance, when you are lying down, it would appear to a person who sees subtle energies that each of your chakras sits upon this foundational energy. Grid energy is sturdy and fundamental. But severe trauma can damage your foundation, and when this occurs, it does not usually repair itself spontaneously. Rather, the other energy systems adjust themselves to the damaged grid, much as a personality may be formed around early traumatic experiences. Repairing a person's basic grid is one of the most advanced and intense forms of energy medicine. If a grid's structure or a car's chassis is sound, you never notice it is there; if it is damaged, nothing else is quite right.

DO YOU WANT YOUR HEALTH CARE APPROACH TO BE BASED ON THE REALITY OF ENERGY OR THE ILLUSION OF MATTER?

Einstein told us that "energy is everything," and modern physics has confirmed, in essence, that the fundamental "stuff" of stuff isn't stuff at all. It is energy, somehow constellated into a form that we *experience* as solid. You might think that would hold some implications for the way we approach healing. But it makes no intuitive sense, so we simply discount it. I mean, is the material world simply an illusion?

Actually, yes. For starters, the atoms of which you are composed are one part nucleus and, depending on the atom, 10,000 to 100,000 parts space. That is, they are almost all space. The atom's empty space is encased by "particles" called electrons that surround it something like a cloud, but they aren't really particles at all. They are mysteries that seem to function sometimes like particles and sometimes like waves, more like light. In fact, according to some physicists, a proton is composed of light circling a point, an electron is light moving between two points, and *those* are the building blocks of matter. Beyond that, if you increased the nucleus of an atom to the size of a billiard ball, the nearest electron would be more than a mile away. In between the nucleus and the electron is space. If atoms are the fundamental building blocks of the chair or couch on which you are sitting, and if space is the most fundamental property of each of its atoms, how does it support you? If you are almost all space, your clothes are almost all space, and the couch is almost all space, why don't you simply pass through one another?

At the most basic level, it would certainly seem that if two atoms are headed on a collision course they would—as minuscule galaxies of mostly space—indeed pass through one another. But they do not. The reason is their *energetic* charge. Because the electron "cloud" surrounding each atom carries a negative charge, the atoms repel one another.[12] At the larger level (no offense) of your bottom on the couch, your electrons and its electrons oppose one another, so you are, to a degree, actually *levitating* away from the couch.[13]

The more you look at the fundamental building blocks of matter, in fact, the more puzzling everything becomes. Niels Bohr, widely considered one of the greatest physicists of the twentieth century, once quipped that "if you think you can talk about quantum theory without feeling dizzy, you haven't understood the first thing about it." Returning to the question of why we don't experience everything as energy if it really is energy, there may be much more than our senses perceive. As my dear friend Ann Mortifee's son, Devon, observed when he was eight: "If I were a dog, it would all be in black and white. If I were a mosquito, I would only see heat waves. If I were a snake, it would all be infrared. So I guess you can never really know what is there; it all depends on whose eyes are looking." Through my eyes, at least, the human body is a system of living energy, and energy medicine appears to be barking up the right tree.

GENES OR FIELDS?

Western medicine tends to wait for broken parts and broken systems and then tries to fix them. Energy medicine keeps a focus on your body's overall functioning and is as much for the prevention of illness as it is for treating it. Conventional medicine is long on invasive interventions and short on prevention.

The prevailing paradigm you learned in high school biology is that the blueprint for building and maintaining your body is coded in your genes. But that is only part of the story. Deep in the nucleus of each cell of your body are indeed, as you were taught, 46 chromosomes (23 from each parent), threadlike structures made of the nucleic acid DNA. Your chromosomes carry close to 24,000 genes,[14] the fundamental units of heredity by which characteristics such as facial structure, hair color, eye color, height, build, introversion/extroversion, and specific types of intelligence, as well as susceptibility to certain diseases, are passed on from the parents to the offspring.

How all this is actually coordinated is a mystery. We do know that genes instruct cells to produce proteins or other molecules. However, each cell undergoes some 100,000 chemical reactions per second, and these are exquisitely coordinated with the actions of many of the body's other 100 trillion or so cells. The number of sensors and switching devices that would be required in each cell to orchestrate such a lavish operation is of an order that far exceeds any known mechanisms.

And how does the gene, which gives the instructions, even know it is part of, say, a kidney cell and not part of a liver cell? The chromosomes and genes in the nucleus of every cell are identical. In fact, when primitive, undifferentiated tissue cells from a salamander were grafted near the tail, they grew into another tail; when grafted near the hind leg, they grew into another leg.[15] These identical genes, depending on where they are located, give their instructions as if they are perfectly aware of not only what is going on all over the body but also what is needed from them in relationship to it all. How do genes know which instructions are required? Lynne McTaggart asks, "If all these genes are working together like some unimaginably big orchestra, who or what is the conductor?"[16]

Western science has no idea. No one has identified chemical mechanisms that inform the genes about the state of the entire body. What of the concept of an "energy field" where the information is more or less "broadcast" to the genes? It is hardly plausible that billions of as-yet-undiscovered chemical reactions would be carrying out the job. Think instead of a TV signal activating the pixels of your television and the coil of its speakers to produce the next episode of *Boston Legal*. It is quite plausible that some sort of field is broadcasting information and coordinating cell behavior. Wild speculation? Perhaps. But until we have scientific instruments that can reliably detect such fields, healers who are sensitive to the body's subtle energies are themselves just such instruments. And by influencing these fields, as you will learn to do in the coming pages, you will be able to impact the moment-to-moment expression of the genes that are instructing your body with its every breath, every heartbeat, and every brain wave. An incredible universe is spinning right there beneath your skin. Interventions into the energy fields that coordinate that universe have some dramatic advantages over drugs and surgery.

FIELDS HOLD INFORMATION

While biologists have carefully mapped the actions of homeostasis and other complex feedback mechanisms, these are the notes, not the melody. No chemical explanations account for how the whole scheme is managed. The intelligence displayed by the body's energies and energy fields, however, is stunning, and may provide the explanation. Even at the subatomic level, as Einstein observed, "the field is the sole governing agency of the particle."[17]

The idea that energy fields impact biological development keeps reemerging within Western science. Sir Isaac Newton provided the first modern description of the aura in 1729 when he wrote of an "electromagnetic light, a subtle, vibrating, electric and elastic medium that was excitable and exhibited phenomena such as repulsion, attraction, sensation, and motion." In the 1930s Harold Burr, a neuroanatomist at Yale, measured the electromagnetic field around an unfertilized salamander egg and found that it was shaped like a mature salamander, as if the blueprint for the adult were already there in the egg's energy field.[18] He was amazed, actually, to find that the electrical axis that would later align with the brain and spinal cord was already present in the unfertilized egg. He went on to find electromagnetic fields surrounding all manner of organisms, from molds to plants to frogs to humans, and he was able to distinguish electrical patterns that corresponded with health and with illness.

The role of electromagnetic fields in healing is well established.[19] When an animal is injured, for instance, electrical currents connecting enormous numbers of cells are produced as part of the growth and repair mechanism. In addition to such internally generated fields, when external currents are applied to an area of tissue, large numbers of cells act in concert. They initiate physiological changes, which may be for better or for worse. This finding may explain the therapeutic effects when a healing practitioner's hand (which itself generates a measurable electromagnetic field) has been held in the proximity of diseased or injured tissue.[20] When a healer is at rest and then begins to work with a patient, the electromagnetic field of his or her hands increases significantly.[21]

While the idea that fields carry information is still in many ways unfamiliar in our culture, powerful examples have made their way into the public eye. Some of the most dramatic instances have been the heart transplant patients who have information about the person whose heart now beats in their bosom. No explanation makes sense other than that the heart carries its own energy field (indeed, the electromagnetic field of the heart is about 60 times greater in amplitude than that of the brain, and its magnetic field according to some estimates is up to 5,000 times stronger[22]) *and* that this field carries information about the person. Consider the following story, told by a psychiatrist to an international group of psychotherapists, about one of her patients:

> *I have a patient, an eight-year-old little girl who received the heart of a murdered ten-year-old girl. Her mother brought her to me when she started screaming at night about her dreams of the man who had murdered her donor. She said her daughter knew who it was. After several sessions, I just could not deny the reality of what this child was telling me. Her mother and I finally decided to call the police and, using the descriptions from the little girl, they found the murderer. He was easily convicted with evidence my patient provided. The time, the weapon, the place, the clothes he wore, what the little girl he killed had said to him . . . everything the little heart transplant recipient reported was completely accurate.[23]*

THE PHYSICIST AND THE PHYSICIAN
DRINK FROM DIFFERENT CUPS

This account, which is consistent with the experiences of many organ recipients, seems to beg for an explanation that transcends conventional understanding of the material world. The great irony here is that the paradigm or worldview embraced by Western medicine is a century behind the paradigm used by modern physics. In 1905, Albert Einstein published his piercing formula, $e=mc^2$, showing that energy and matter are interchangeable. This discovery revealed that Newtonian physics, which focuses on the mechanics of life and is the basis of Western medicine, gives us only a glimpse of a much larger story. The implications of this larger story burst into our collective psyche on August 6, 1945, when the myth of Prometheus, who stole fire from the gods, became the terrifying reality of a humanity that suddenly had an atomic bomb. But the realization that the billiard ball–like atoms of a century ago are really packets of energy—unique in their distribution of positive and negative charges, spin rate, and vibrational pattern[24]—is also about to revolutionize some of our most cherished Promethean inventions, such as televisions, cell phones, and computers, all originally based on electromagnetic effects.

Nonetheless, Western medicine continues to focus on the physiology and chemistry of the body rather than its energies, preserving the hallowed place of pharmaceuticals and surgery over energy treatments in our health care practices. But leading-edge science does not support this unilateral approach. According to cell biologist Bruce Lipton, who has served as a research scientist on the faculty of Stanford Medical School, hundreds upon hundreds of scientific studies over the past fifty years have revealed that "every facet of biological regulation" is profoundly impacted by the "invisible forces" of the electromagnetic spectrum. He explains that specific patterns of "electromagnetic radiation regulate DNA, RNA, and protein synthesis, alter protein shape and function, and control gene regulation, cell division, cell differentiation, morphogenesis [the process by which cells assemble into organs and tissues], hormone secretion, nerve growth and function," essentially the fundamental processes that contribute to "the unfolding of life." But, he laments, "though these research studies have been published in some of the most respected mainstream biomedical journals, their revolutionary findings have not been incorporated into our medical school curriculum."[25]

What does this disregard of the role of energy in regulating biological processes mean for you? It means that procedures are more invasive and less precise than energy medicine. Electromagnetic imbalances cause the body to produce chemicals, such as progesterone or estrogen, to restore balance. Those chemicals are produced in the precise quantities needed and are directed only where needed. When medications are introduced into the bloodstream to do the same thing, their dosage is based on averages and guesswork. They can travel to and impact parts of the body that are not intended, with disastrous results, including increases in heart disease, strokes, and breast cancer among women who have undergone hormone replacement therapy. Although these are blandly called "side effects," between 100,000 and 300,000 people in the United States die each year from medications taken as prescribed, and unintended effects of medical treatment are by some estimates our leading cause of death.[26] A team that reviewed government health statistics over the past decade concluded, "When the number one killer in a society is the health care system, then that system has no excuse except to address its own urgent shortcomings . . . beginning at its very foundations."[27]

CHEMICAL SIGNALS AND ELECTRICAL SIGNALS: THE TORTOISE AND THE HARE

While the chemical paradigm remains at the foundation of conventional medicine, the energy paradigm is gaining ground for good reason. Electromagnetic frequencies are vastly more efficient than chemical signals (moving at 186,000 miles per second versus less than a centimeter per second). Plus most of the information in chemical signals is lost because so much of the operation goes into simply making and breaking chemical bonds.

Energy fields, on the other hand, can evolve and quickly adapt, keeping up with rapid and unending changes in the physical body, the environment, and other energy fields. Lipton summarizes the benefits and costs of energy versus drug treatments: "Energy signals are 100 times more efficient and infinitely faster than physical chemical signaling. What kind of signaling would your trillion-celled community prefer? Do the math!"[28]

Conventional medical treatments still do not for the most part take advantage of the powerful ways energy can transmit information in biological systems (with some notable exceptions, such as heart pacemakers, brain "pacemakers" for treating Parkinson's disease and depression, harmonic frequencies that dissolve kidney stones, and the use of magnets for alleviating tendinitis, facial paralysis, and optic-nerve atrophy). In another irony, conventional medicine has had no difficulty accepting diagnostic instruments that are based on the concept of energy *as information*. Healthy and unhealthy tissues have different electromagnetic properties that can be detected in scanned images. Energy-scanning devices analyze the frequencies in these tissues. MRIs, EEGs, EMGs, and CAT scans have proven their ability to noninvasively detect illness.

Lipton observes that "diseased tissue emits its own unique energy signature, which differs from the energy emitted by surrounding healthy cells,"[29] and he goes on to speculate that scientific evidence suggests that we will be able to tailor energy and waveforms to act as therapeutic agents "in much the same way that we now modulate chemical structures with drugs."[30] Meanwhile, indigenous healers have been making diagnoses based on the unique "energy signatures" of diseased tissues and devising remedies by working with the body's energies going back, undoubtedly, to our earliest times. The re-embrace of their skills and perspective, combined with the almost miraculous power of medical technology, brings us to the threshold of a brave new world in health care.

But even without the technology, the perspective and hands-on skills offered by energy medicine can be a profound and life-enhancing resource for you, and they are available *now*. With these concepts and techniques, you can better manage your energies, your hormones, your mood, and your health. This book shows you how. In the following chapter, we begin with some of my favorite ways for putting this into practice.

Energy Techniques for Health and Vitality[1]

Your body is not just a collection of physical and chemical events.
Like all living systems, it is a unified collection of energy fields.
Take action to alter the quality of these fields, and you can
change the way your body functions for ill or good.

—LESLIE KENTON
Passage to Power

This chapter presents a series of exercises based on six principles:

1. Stretching makes space inside your body for your energies to move in their most natural flow.
2. Clearing toxins supports the healthy flow of your body's energies.
3. A set of simple exercises done on a daily basis can help establish and maintain positive "energy habits" that optimize your health and vitality.
4. It is possible to assess how the energies are flowing in various parts of your body, and to take simple actions that enhance the flow of energies that are stagnant, blocked, or otherwise distressed.
5. Healthy energies move in crossover patterns.
6. You can reprogram deep-seated energetic patterns in your body and mind.

MAKING SPACE FOR YOUR ENERGIES TO MOVE

For your cells and organs to serve you as they are designed to, the energies that sustain them must be given *space to move*. This is a basic physical principle of energy medicine. Stretching is one of the most natural ways to "make space," which is in turn one of the best ways to support the natural rhythms and flows of your body's energies. From watching cats and dogs stretch upon waking to practicing disciplines that have made stretch into a science (such as yoga and tai chi), many models are available.

A BASIC STRETCH TO KEEP YOUR ENERGIES FLOWING

A simple exercise that I call "Connecting Heaven and Earth" is easy to do and is an excellent way to help keep your energies in a good flow. Versions of it have been found in numerous cultures throughout history. I've seen it on Egyptian hieroglyphics in the Museum of London, and variations of it can be found in qigong, yoga, and other disciplines. It is formulated here to also help integrate the left- and right-brain hemispheres and activate the radiant circuits, which carry the healing energies that you experience as joy. If I have a long day with many clients, this is the exercise I am most likely to use between clients to release stagnant energies in myself or any energies I may have taken on from my clients. It also provides a nice meditative break and is refreshing whenever you use it.

Connecting Heaven and Earth (time—about two minutes)

(Please do not be put off by all the words in the instructions here and throughout this book. It is actually easier to do these exercises than to read about them, and the photos will also help.)

1. Rub your hands together and shake them off.
2. Place your hands on the fronts of your thighs with your fingers spread.

3. With a deep inhalation, circle your arms out to your sides.

4. On the exhalation, bring your hands together in front of your chest in a prayer position.

5. Again with a deep inhalation, separate your arms, stretching one high above your head and flattening your hand back, as if pushing something above you. Stretch the other arm down, again flattening your hand as if pushing something toward the earth. Look up to the heavens (see Figure 2-1). Stay in this position for as long as is comfortable.

6. Release your breath through your mouth, returning your hands to the prayer position in front of your heart.

7. Repeat, switching the arm that raises and the arm that lowers. After this first set, do two or more sets.

8. Coming out of this pose the final time, drop your arms and allow your body to fold over at the waist. Hang there with your knees slightly bent as you take two deep breaths. Slowly return to a standing position with a backward roll of the shoulders.

Figure 2-1
CONNECTING HEAVEN
AND EARTH

MOVING TOXINS OUT OF YOUR BODY

Clearing the body of toxins is as important for personal health as removing garbage from local neighborhoods is for keeping a city clean and robust. Many illnesses can be traced to toxic buildup, which also depletes your vitality and darkens your sense of well-being. Organs such as the liver, kidneys, bladder, and intestines work in harmony to rid the body of toxins. The liver itself is a chemical-processing plant that carries out some 100 functions, including filtering toxins and waste products from the blood, producing chemicals that break down fat, and creating urea (the main substance in urine). A powerful, simple way for keeping your liver strong and clean while also supporting the flow of hormones through your bloodstream is to massage points on the hands and feet that stimulate liver function. Breathe deeply throughout the massage.

In this and all the subsequent procedures in this book, unless otherwise noted, when you are invited to "breathe deeply," I suggest you breathe in slowly through your nose and then out through your mouth. This breathing pattern connects the energies of the central and governing meridians at the back of your throat, creating a healing force field that permeates and surrounds your body.

Massaging Your Hands to Keep Your Liver Clear
(time—about a minute)

Figure 2-2
HAND MASSAGE FOR LIVER

1. With your right thumb on top of the fleshy webbed area between the thumb and index finger of your left hand, and your right index and middle fingers on the bottom of the left hand, massage the webbed area. This will stimulate the fourth acupressure point on the large intestine meridian, or the Hoku point, one of the most widely used points in acupressure and acupuncture. If you are pregnant, however, skip this step.[2]

2. Bring your right thumb to the palm side of your left hand and massage up the palm and off each finger.

3. End by bending your fingers backward (see Figure 2-2). Stretching these ligaments releases stagnant energy from the liver meridian (which governs the ligaments) and also frees energies from six other meridians that travel through your hands.

4. Repeat on your right hand.

Massaging Your Feet to Keep Your Liver Clear

(time—about a minute)

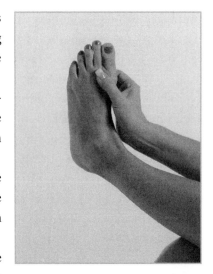

1. Massage the foot point that corresponds with the Hoku point, at the *V* where the big toe meets the second toe on the top of the foot (see Figure 2-3).

2. Use your thumb to massage the area between the tendons of the two toes, while using your fingers to massage the bottom of your foot.

3. It also feels great to massage and clear the "gaite reflexes," the areas on the top of the foot that are between the tendons of each of the toes.

4. Repeat on the other side. When you are done, place your hands around the bottoms of each foot and squeeze the sides hard.

Figure 2-3
FOOT MASSAGE FOR LIVER

I find myself massaging my hands almost every day, and you can readily massage the points on your feet whenever you take a bath, get in a hot tub, or sink into bed. This is a great form of self-care that not only helps your liver function better, it also keeps your hands and feet—which function as "energy antennae"—open and clear.

THE LYMPH SYSTEM IN ENERGY MEDICINE

You know lymph as the clear fluid that comes out of a cut. It is there to remove foreign matter and fight germs. Lymph plays a key role in your immune system by helping counter conditions ranging from colds to cancer. It creates antibodies and produces lymphocytes, specialized white blood cells produced in the lymph nodes, found in your neck, armpits, abdomen, and groin. Your lymph system also carries proteins, hormones, and fats to the cells, and eliminates dead tissue and other waste products. Its role in eliminating toxins, as do organs such as the liver and kidneys, is particularly important in energy medicine.

Because toxins block the flow of energy—and this includes "energy" toxins such as emotional residue, stagnant energies, electromagnetic pollution, and radiation, as well as chemical toxins—the necessity for *moving toxins out of your body* is a second basic principle of energy medicine, along with the need for *space* and *stretch*. The lymphatic system moves both chemical and energy toxins out of the body. Chemical toxins, in fact, are a cause of stagnant energy.

Two types of liquid circulate through your body: blood and lymph, with the lymphatic system sometimes being called the body's "other" circulatory system. You, in fact, have twice as many lymph vessels as blood vessels, but while your circulatory system has your heart to pump your blood, your lymphatic system has no pump. Your body's movement during your daily activities, and during exercise in particular, fosters the circulation of your lymph. But sometimes the accumulation of toxins will cause blockages in the flow of the lymph so it becomes less and less effective in clearing toxins from certain parts of the body.

Reflex points that stimulate the flow of lymph are located on various points throughout your body, particularly on the chest, back, and upper portions of the legs. When they are massaged, the lymphatic system is stimulated and toxins are removed more quickly and effectively. This use was developed in the 1930s by Dr. Frank Chapman, D.O., and is a mainstay in the field of applied kinesiology.[3] These points, called "neurolymphatic reflexes,"[4] have, in my experience, proven to be extraordinarily beneficial in releasing the body's toxins as well as in freeing the flow of energy throughout the body.

David used to question me about whether working these points was really part of energy medicine. He thought they were part of massage. Really, David, who cares! But I mention this because a few influential legislators have too often been persuaded by a professional group that they should create legislation that protects the narrow interests of the professional group. Some states have actually made it illegal for certain health care providers to utilize isolated massage techniques unless they have a massage license. This does not serve the public in any way, shape, or manner. Massage techniques always have been, and always will be, part of many approaches to healing, including energy medicine, and it is not necessary to become a fully trained massage therapist (which I happen to be) to utilize specific techniques effectively and appropriately. Massaging neurolymphatic reflex points releases toxins into the lymphatic vessels and subsequently into the bloodstream so that they can be eliminated. Clearing toxins in this way makes more space for energy to flow, and that is why working with these points is an important technique within energy medicine.

There are 90-some neurolymphatic reflex points on the surface of the body, sometimes simply referred to as "lymphatic points," though, as reflexes, they are not necessarily situated directly over the lymph nodes, vessels, or other lymph tissue. Figure 2-4 gives a map. When the neurolymphatic reflex points become clogged, every system in your body is compromised. Congested neurolymphatic reflex points feel sore when massaged. For that reason, they are not hard to locate. And there are so many of these points so close to one another that you won't miss them. Massaging them is a way to clear them and allow the energy that has been blocked to flow again. If you simply start to push in on these points with your middle finger, using some pressure, you are likely to quickly identify at least a few that are sore. If you are certain there is no injury, no strain from exercise, and no medical condition responsible for the soreness, the chances are good that you have located a neurolymphatic reflex point needing attention.

The Lymphatic Massage (time—about 10 seconds on each point)

1. With your middle finger pressing neurolymphatic reflex points on your chest, identify one that is sore.
2. Press into the skin over this spot with two or three fingers and massage by firmly moving the skin in all directions.
3. Press hard enough that you really feel the pressure but not so hard that you risk bruising yourself. Massage each sore point for about ten seconds.

Massaging your lymph points is a simple procedure that will enhance the flow of your energies while also supporting your hormones to function as nature intended. A few cautions with this method need to be followed. You should not massage painful points if the pain is based on a bruise or injury. Be conservative about massaging too many points that are sore in the same session. Since massaging these points does release toxins into the lymph and then the bloodstream, you do not want to overwhelm your body's capacity to eliminate them. This caution applies particularly if you are or have recently been ill. It also applies if you have an autoimmune disorder such as multiple sclerosis or Parkinson's disease, since it is

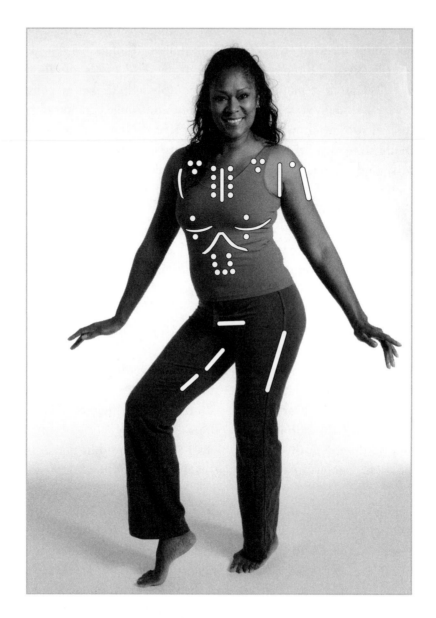

Figure 2-4
NEUROLYMPHATIC REFLEX POINTS

sometimes difficult for people with autoimmune disorders to assimilate changes in the body's chemistry; so go slow. But for the vast majority of situations, these are wonderful points to massage. The process is generally pleasant (in the sense of "it hurts so good"), you will often feel the pleasing sensation of energy moving in the area being cleared immediately after the self-massage, and it is an easy way to help keep your energies flowing. So with the Connecting Heaven and Earth exercise, the Hand Massage, the Foot Massage, and the Lymphatic Massage, you have four simple ways to maintain your body as a friendlier container for your energies.

A FEEL-GOOD DAILY ENERGY ROUTINE

Some of the techniques you will learn in this book, particularly those presented in this chapter, will help you keep your overall energies vital and balanced. They also have the effect of maintaining harmony among the energies that govern your body's chemistry. Other procedures, presented in each of the subsequent chapters, are oriented toward working with specific conditions such as PMS, fertility problems, or hot flashes. Energy medicine often begins, however, with a focus on the fundamental energies that affect general well-being. Following is a set of powerful techniques, presented as a five-minute routine, which I encourage you to do every day.

After working with more than 10,000 clients in 90-minute individual sessions, giving most of them back-home assignments and seeing what occurred, I have some ideas about which techniques are the most potent and which will have a positive impact for just about everyone. When we were writing my first book, *Energy Medicine,* we wanted to give people a simple routine they could use every day to keep their "energies humming." I identified the procedures that had the greatest impact on the greatest number of people and selected those that, as a set, tended to bring all of the body's energy systems into balance.

The ultimate effect is to reestablish positive "energy habits." As we adjust to the stresses and unnatural practices and substances that characterize modern life, our energy systems make compromises as they valiantly attempt to keep us healthy. Often, however, these compromises throw our energies into habitual patterns that hurt us more than help us. The Five-Minute Daily Energy Routine is like pressing the reset button, helping restore your body's natural energy flows.

The Daily Energy Routine follows. I refer to it throughout this book, and I suggest that you practice it regularly. All the subsequent techniques are based on the assumption that you are taking basic steps to keep your overall energies in balance. In the same sense that a woman with blurry vision who is trying to improve her reading ability might consider eye exercises before taking a speed-reading class, some basic measures are necessary before fine-tuning will be as effective.

I know it is no small thing to suggest that you build another routine into your life. We are all extraordinarily busy, an epidemic of modern life. But some investments pay off. I promise you that the Daily Energy Routine, practiced regularly, will give you a good return—in terms of how you feel and function—on the small daily investment it requires.

Beyond my personal promise, however, some very encouraging research supports this claim. One of the practitioners I have trained, Linda Geronilla, Ph.D., a psychologist as well as energy medicine practitioner, taught a group of 18 school-teachers a modified version of the Daily Energy Routine and studied whether their brain patterns changed. Certain activities have been shown to optimize specific areas of the brain. Dr. Geronilla patterned her study after the work of Daniel Amen, M.D., who has investigated the relationship of brain activity and optimal functioning with more than 30,000 subjects.[5] At the end of eight weeks of daily practice, positive changes could be inferred in four of the six brain areas that Dr. Amen has identified as most important to target in programs for optimizing people's performance (the prefrontal cortex, the cerebellum, the temporal lobe, and the anterior cingulate gyrus). The participants also reported improvements in capacities such as memory, concentration, and energy level. A control group did not report such improvements and did not show significant changes in any of the six brain areas over the eight-week period. These findings corroborate Dr. Amen's previously untested but published prediction that "Brain Gym" (an approach that uses similar exercises for working with the body's energies) will enhance the cerebellum— helping people think more clearly and quickly—and improve their "judgment, attention, and overall brain health."[6]

While this should persuade you to seriously consider making the Five-Minute Daily Energy Routine a habit, I also know from experience that you will be more likely to maintain a program like this if you tie it into an activity you already do. If you exercise regularly, or do yoga, tai chi, or Pilates, it can be a great warm-up or cooldown. If you meditate, it can bring you into a centered space from which you can go deeper. Some people, particularly those who are not morning persons, do it before they get out of bed. Some people do it as a kind of transition ritual when they come home from work. Some people do it as part of their bath or shower. The more comfortable you are, the better. It is not a matter of "no pain, no gain"—just the opposite, in fact. The Five-Minute Daily Energy Routine consists of the Three Thumps, the Cross Crawl, the Wayne Cook Posture, the Crown Pull, the Lymphatic Massage, the Zip-up, and the Hook-up.

Three Thumps (time—about 30 seconds)

Certain points on your body, when tapped with your fingers, will impact your energy field in predictable ways, sending electrochemical impulses to your brain and releasing neurotransmitters. By tapping three specific sets of points, a technique I call the "Three Thumps," you can activate a sequence of responses that will restore you when you are tired, increase your vitality, and keep your immune system strong amid stress. Do not be too concerned about finding the precise location of each point. If you use several fingers to tap in the vicinity described, you will hit the right spot.

Thump #1: Your K-27 Points. Acupuncture points (also called acupressure points with procedures in which needles are not used) are tiny energy centers arranged along the body's fourteen major meridians or energy pathways. The K-27 points (the paired 27th acupressure points on the left side and the right side of the kidney meridian) are juncture points that affect all of the other meridians. Tapping or massaging your K-27 points also sends signals to your brain to adjust your energies so you will be more alert and able to perform more effectively. Tapping these points can energize you when you are feeling drowsy and focus you if you are having difficulty concentrating.

To locate these points, place the pointer finger of each hand on your collarbone and move your hands toward each other until you reach the two inside corners of your collarbone. Drop straight down from these points to about an inch below your collarbone. For most people, there is a soft spot or indentation there. Then breathe

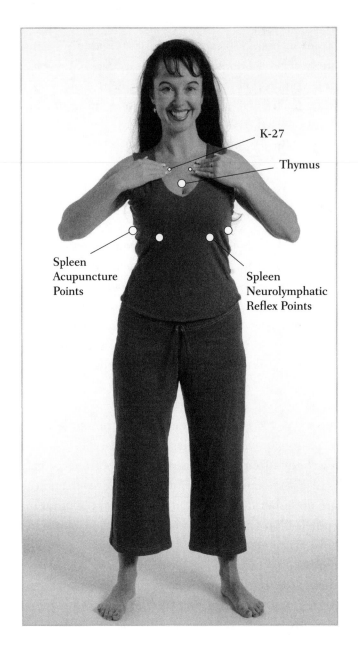

Figure 2-5
K-27, THYMUS, AND SPLEEN THUMPS

K-27

Thymus

Spleen
Acupuncture
Points

Spleen
Neurolymphatic
Reflex Points

slowly and deeply as you firmly tap or massage your K-27 points (see Figure 2-5) for two or three deep breaths.

Thump #2: Tarzan Thump. Tapping the area over the thymus gland (think of Tarzan) is a simple technique that awakens the body's energies, sharpens the immune system, and increases your strength and vitality. The thymus thump can help you if you are feeling bombarded by negative energies, catching a cold, or fighting an infection, or if your immune system is otherwise challenged. Your thymus gland supports your immune system. If you keep making choices that are not in harmony with your body's needs and design, your thymus's surveillance mechanism becomes confused. Thumping the thymus helps stimulate and reset it:

1. Place the fingers of either or both hands in the center of your sternum, at the thymus gland, about two inches below the K-27 points in the center of your chest (see Figure 2-5).
2. Using your thumb and all your fingers as you tap firmly, breathe slowly and deeply, in through your nose and out through your mouth for two or three deep breaths.

Thump #3: Spleen Points. Tapping points that affect the spleen meridian is a quick way to lift your energy level, balance your blood sugar levels, and bolster your immune system.

1. Tap the *neurolymphatic* spleen points firmly for about 15 seconds (see Figure 2-5). The *neurolymphatic reflex points* are located beneath the breasts in line with the nipples and down one rib.
   Alternate: Tap the spleen *acupuncture* points firmly for about 15 seconds. These acupuncture points are on the sides of the body about four inches beneath the armpits.
2. If either set is more tender, tap those points.
3. Breathe slowly and deeply, in through your nose and out through your mouth, as you tap.
4. Tap (or massage) firmly while you take two or three deep breaths.

Cross Crawl (time—about 30 seconds)

The cross crawl facilitates the crossover of energy between the brain's right and left hemispheres. It will
help you feel more vital, think more clearly, and move
with better coordination. It is essentially an exaggerated walk. Most people find it energizing. If you find
that it drains you or tires you, instead do the Homolateral Crossover (page 67). To do the Cross Crawl:

Figure 2-6
CROSS CRAWL

1. First briskly tap the K-27 points (see page 53)
   to assure that your meridians are moving in a
   forward direction.
2. The cross crawl is as simple as marching in
   place. While standing, lift your right arm and
   left leg simultaneously (see Figure 2-6).
3. As you let them down, raise your left arm and
   right leg. If you are unable to do this because
   of a physical disability, here is an alternative: while sitting, lift your right
   knee and touch it with your left hand, then lower it and lift your left knee
   and touch it with your right hand.
4. Repeat, this time exaggerating the lift of your leg and the swing of your
   arm across the midline to the opposite side of your body.
5. Continue in this exaggerated march for at least a minute, again breathing
   deeply in through your nose and out through your mouth.

Wayne Cook Posture (time—about 90 seconds)

I use the Wayne Cook Posture when I am overwhelmed or hysterical, cannot get
clarity on a situation, cannot concentrate, must confront someone, or am upset after someone has confronted me. This procedure is named to honor Wayne Cook, a
pioneering researcher of bioenergetic force fields, who invented the approach that
I have modified into the form presented here. Perhaps more than any other single
approach I teach, the Wayne Cook Posture can calm you, bring order to your
thinking, and help you better understand and confront the problems you face.

This technique is effective even when the upset is so intense that you are unable to quit crying, are finding yourself snapping or yelling at others, are sinking into despair, or are feeling that you are beyond exhaustion. It helps process stress hormones. Almost immediately, you will begin to feel less crazy and less overwhelmed. To do the Wayne Cook Posture:

1. Sitting in a chair with your spine straight, place your right foot over your left knee. Wrap your left hand around the front of your right ankle and your right hand over the ball of your right foot, with your fingers curled around the side of the right foot (see Figure 2-7a).

2. Breathe in slowly through your nose, letting the breath lift your body as you breathe in. At the same time, pull your leg toward you, creating a stretch. As you exhale, breathe out of your mouth slowly, letting your body relax. Repeat this slow breathing and stretching four or five times.

a   b

Figure 2-7
WAYNE COOK POSTURE

3. Switch to the other foot. Place your left foot over your right knee. Wrap your right hand around the front of your left ankle and your left hand over the ball of your left foot, with your fingers curled around the side of your left foot. Use the same breathing.

4. Uncross your legs and "steeple" your fingertips together so they form a pyramid. Bring your thumbs to rest on your "third eye," just above the bridge of your nose. Breathe slowly and deeply, in through your nose and out through your mouth, about three or four full breaths (see Figure 2-7b).

5. On the last exhalation, curl your fingers into the middle of your forehead and separate them, firmly and pleasantly, pulling across your forehead to your temples.

6. Slowly bring your hands down in front of you. Surrender into your own breathing.

Crown Pull (time—about 30 seconds)

A great deal of energy is processed in your brain and skull, but it can become stagnant if it doesn't release and move out through the energy center at the top of your head (called the crown chakra in yoga tradition). The Crown Pull physically opens this chakra so energy can move through it. It clears the cobwebs from your mind and brings a calm to your nervous system, releasing mental congestion, refreshing your mind, and opening you to higher inspiration. People today are hungry to receive higher forms of inspiration and guidance and to feel more fully connected with the force of creation or God or however they conceive of the larger picture, and the crown chakra is a key to such psychic opening. The Crown Pull is a powerful way to begin to open your crown chakra, and it also has the more mundane benefit of simply clearing congested energies. To do the Crown Pull, breathe deeply, in through your nose, out through your mouth, and:

1. Place your thumbs at your temples on the sides of your head. Curl your fingers and rest your fingertips at the center of your forehead.

2. Slowly, and with some pressure, pull your fingers apart so that you stretch the skin above your eyebrows (see Figure 2-8a).

3. Rest your fingertips at your hairline and repeat the stretch.

a   b   c

Figure 2-8
CROWN PULL

4. Continue this pattern with your fingers curled, pushing in at each of the following locations:
   a. Fingers at the top of your head, with your little fingers at the hairline. Push down with some pressure and pull your hands away from each other, as if pulling your head apart (see Figure 2-8b).
   b. Fingers over the curve at the back of your head, again using the same stretch (see Figure 2-8c).
   c. Fingers at the bottom of your head, again using the same stretch.
   d. Continue, pulling to the sides of your neck with three passes (top of neck, middle of neck, and bottom of neck), finally resting your fingers on your shoulders.
   e. After a deep breath, pull your fingers firmly forward over the tops of your shoulders and let them drop.

Lymphatic Massage (time—at least a minute)

Massaging the neurolymphatic reflex points discussed beginning on page 48 is so valuable that I have placed the procedure in the Daily Energy Routine. Try to work two or three sore spots each day (see Figure 2-4, page 50—there are many to choose from) for about 10 to 20 seconds each. You may find that points that have been chronically sore become less tender, sometimes almost immediately. This indicates that the toxins are being eliminated. An alternative to massaging your own points is to work with a partner and give each other a "Spinal Flush." Because major neurolymphatic reflex points affecting every meridian run along either side of the spine, those points are a great focus for a shared session, and the Spinal Flush is a real treat. Almost every morning when we are not on the road, I make David a health drink. Then he gives me a Spinal Flush. I hope he doesn't read this and decide to renegotiate, but I am getting the better part of the bargain. Instructions for the Spinal Flush:

1. Lie facedown, or stand three or four feet from a wall and lean into it with your hands supporting you. This helps stabilize your body while your partner applies pressure to your back.

2. Have your partner massage the points down both sides of your spine, half an inch to an inch to the left and right of the spine, using the thumbs or middle fingers and applying body weight to get strong pressure. Massage from the bottom of the neck down to the bottom of the sacrum (See Figure 2-9).

3. Have your partner go down the notches between your vertebrae and deeply massage each point. Staying on the point for at least three seconds, your partner moves the skin in a circular motion with strong pressure, checking to make sure that the pressure is comfortable.

4. Upon reaching your sacrum, your partner can repeat the massage or can complete it by "sweeping" the energies down your body. From your shoulders, and with an open hand, your partner sweeps all the way down your legs and off your feet, two or three times.

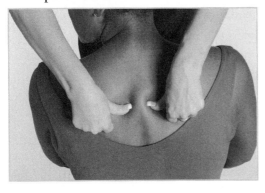

Figure 2-9
SPINAL FLUSH

Do not be concerned about a point being missed. Each of your meridians will be covered by simply going between all the notches. Rather than knowing which meridians are associated with which points, simply ask for special attention to any points that are sore. If you are moving through intense emotional or physical stress, or if you have been exposed to environmental toxins, doing the Lymphatic Massage or the Spinal Flush quickly helps clear your lymphatic system. The Spinal Flush not only cleanses the lymphatic system, it also stimulates the cerebrospinal fluid, clearing your head as well. It is a quick rebalance, and of all the energy techniques I've seen, it probably delivers the most benefit for the least effort in the greatest number of situations. As a cold is coming on, the Spinal Flush can stop it in its tracks. I also regularly recommend it to couples, both as a way of lovingly caring for each other and as a way to head off problems. If you sense that an interchange is headed toward an argument, tell your partner, as lovingly as you can, "Up against the wall!" and firmly work the neurolymphatic reflex points. This simple technique immediately reduces built-up stress and takes the edge off emotional overreactions.

Zip-up (about 20 seconds)

When you are feeling sad or vulnerable, the central meridian, one of the two energy pathways that govern your central nervous system, can be like a radio receiver that channels other people's negative thoughts and energies into you. It's as if you are open and exposed. The central meridian runs like a zipper from your pubic bone up to your bottom lip, and you can use the energies of your hands to "zip it up." The Zip-up will help you feel more confident and positive about yourself, think more clearly, and protect yourself from negative energies that may be around you. Pulling your hands up the central meridian draws energy along the meridian line. Before doing the Zip-up, briskly tap the K-27 points to assure that your meridians are moving in a forward direction. Then:

Figure 2-10
Zɪᴘ-ᴜᴘ

1. Place your hand or hands at your pubic bone, which is the bottom end of the central meridian.

2. Take a deep in-breath as you simultaneously move your hand(s), slowly and with deliberation, straight up the center of your body (see Figure 2-10), to your lower lip, where the meridian ends.

3. By continuing upward, however, bringing your hands past your lips and exuberantly raising them into the sky, you connect your central meridian with your aura and with forces that are beyond you.

4. Turn your palms outward and, slowly releasing your breath, extend your arms out to the sides as far as they will go and back down to your public bone.

5. Repeat three times.

Hook-up (time—15 to 20 seconds)

The Zip-up strengthens the central meridian. The central meridian, which sends energy up the front of your body, works in tandem with the governing meridian, which sends energy up your spine. The two meridians meet at the back of your throat. The Hook-up connects these two meridians, bridging their energies between the front and back of your body and between your head and torso. It increases your coordination and stabilizes all your energies, including the aura energies that surround you. It is one of the most powerful tools I know for quickly centering myself. It has immediate neurological consequences. Several of my students have reported using it with someone who has started to seizure and seeing the seizure stop within 15 to 20 seconds. To do the Hook-up (again, remember to breathe in through your nose and out through your mouth):

Figure 2-11
Hook-up

1. Place the middle finger of one hand on your third eye (between your eyebrows above the bridge of your nose).

2. Place the middle finger of the other hand in your navel.

3. Gently press each finger into your skin, pull it upward, and hold for 15 to 20 seconds (see Figure 2-11).

ASSESSING YOUR BODY'S ENERGIES

Some methods in energy medicine, such as those in the Five-Minute Daily Energy Routine, can benefit virtually anyone. Other procedures are tailored to a person's unique energy patterns. Clues indicating what your body needs energetically can be found simply by noticing what you feel when you tune in to your inner sensations and identifying areas of discomfort, by watching the shifts in your emotional landscape, or by noting physical symptoms. Symptoms are clues about underlying energy disturbances. In addition to observing what is going on in your body and mind, a technique for assessing the body's energies can be invaluable in your health care choices.

You can "energy test" to assess the state of your body's energies and to find where disturbances exist in their flow. This can lead to highly specific information. For example, your liver meridian may be depleted of energy, your kidney meridian overcharged, or your third chakra sluggish. You can also energy test to check whether a particular remedy (conventional or unconventional) is working, or at least what impact it is having on your body's energies.

Energy testing comes from the field of Applied Kinesiology, which uses the term "muscle testing" for the procedure. Applied Kinesiology was developed in the 1960s by chiropractor George Goodheart. Goodheart identified relationships among muscles, organs, and the body's *meridians,* or "subtle energy pathways." At least 670 acupressure points are distributed over the body's 14 meridians. The needling of an acupuncture point (or massaging of an acupressure point) affects the energy flow of the meridian on which it is located. Goodheart was able to test and verify these relationships by drawing upon the field of kinesiology (the study of muscles and how muscles coordinate to move the body). Kinesiologists had developed muscle-testing procedures for evaluating isolated muscles according to their relative strength and range of motion.[7]

Since each major muscle is associated with a meridian, Goodheart reasoned, weakness or restricted range of motion in an uninjured muscle indicates impairment in the meridian or energy flow through that muscle. By implication, the organs that receive energy from that meridian might also be affected. Applied Kinesiology was a bridge from traditional Chinese medicine into Western culture, and it has had a strong influence on many of the contemporary approaches to energy medicine, including my own. It utilizes a variety of methods—including holding, massaging, and tapping specific acupressure points—to correct impaired energy flow in meridians that have been identified by a muscle test.

While there has been growing scientific support for the validity of energy testing,[8] the procedure is as much art as science. Energy testing is done in many ways by different practitioners, and there are enough variables and nuances in a single energy test that teaching people how to do the procedure, as well as conducting research on it, is fraught with dilemmas.

I detailed the procedure in my previous book, and I have introduced it to tens of thousands of people in my classes. While research has shown a strong correlation between the results found by people new to energy testing and highly experienced practitioners,[9] I always give the caveat that it takes considerable practice to become proficient with the method, to know how to control all the variables, and to be confident in the results. The mind, for instance, is one of the factors that can interfere with an accurate test. Simply holding a thought shifts your energies. If you are longing for a hot fudge sundae and you put it into your energy field and test the flow of a meridian, you may be testing the effect of your anticipation of the good taste rather than the effect the sugar will have on your energy field. If you do not know how to control for the effects of the mind, the test may show what you want or expect instead of what is.

I have adapted the instructions for energy testing from *Energy Medicine* and placed them in the Appendix. I put them in the Appendix because I do not want to be overly reliant on energy testing in this book. It is a skill that is invaluable and that you can use with great benefit for the rest of your life. But it is best to receive live, hands-on training at first, or to consult a professional. There is, unfortunately, much nonsense and misapplication of the method, so that it has sometimes been reduced to little more than a parlor trick. But its potential is enormous. I have many times been consulted by physicians for help in determining the proper type and amount of medication for a tricky case, including having being brought several times into an operating room to help with delicate decisions about anesthesia.

I have designed this book so it can be used without energy testing, but I have also indicated places where you can benefit from energy testing, either by having learned it yourself or by engaging a qualified practitioner. Energy testing is the most effective method I know for helping people who don't "see" energy work with the body's energies as if they did see it. It is invaluable in deciding how to apply energy medicine procedures and to track their effects. So I encourage you to use the instructions in the Appendix as a tutorial for developing this skill.

And even before you have become proficient with the method, a great benefit of learning energy testing beyond being able to do specific assessments is the way you begin to attune yourself to the world of energies in which we all live. Being in harmony with your energies expands your connection to life, and energy testing gives you a way to affirm your experiences with the energies within and around you. This also opens you to other dimensions. When you attune yourself to the energy realm, you begin to sense that there is a grander purpose to your existence. From finding answers to daily health concerns to exploring the mysteries of the universe, energy testing is a wonderful tool that you can use all your life. You will also have an opportunity to do some optional experimentation with energy testing in the following section if you are already adept in the technique or choose to take a detour to the Appendix now to begin studying energy testing.

YOUR BODY'S VITAL CROSSOVER PATTERNS

Energies need to cross over freely throughout the body. The crossover pattern that I call the Celtic Weave is essential for maintaining the health of your energies and that of your body. I've had two class participants, both male nurses, one from Brazil and one from Nepal, who immediately recognized this energy system when I described it. Each was expected by the hospital where he worked to regularly "Celtic Weave" his patients, one in a burn unit, the other in a general ward, though different names were used for the procedure.

The Celtic weave is both an independent energy system within the body, akin in that sense to the meridians, chakras, and aura, and also a procedure for supporting the body's crossover patterns. As an *energy system*, the Celtic weave laces through all of your other energy systems and creates a resonance among them. It is the weaver of your force fields. It holds your entire energetic structure together. As a *procedure*, the Celtic Weave connects all of your energies so they operate as a single web. Touch one strand, anywhere, and your whole system reverberates in harmony.

The Celtic Weave, both as an energy system and as a procedure, draws all the body's energy systems together into a tight web of communication so that information can easily travel wherever it needs to go. This enables your energy systems to work in harmony and cooperation (the energy systems, as opposed to the Celtic Weave technique). When the Celtic weave is dynamically engaged, you have a sense of power, a feeling of being charged, and your energies start humming. I've known several energy healers who can *hear* the Celtic weave in their clients, particularly when the person is healthy and the pattern is distinct. They say it literally sounds like a musical hum.

Everyone's energy is unique, yet the Celtic weave is universal. It crosses the energies over from each hemisphere of the brain to the opposite side of the body. Its largest formation is a figure eight or infinity sign extending from head to toe. In the caduceus, in the curved lines that cross seven times as used in yoga tradition, and in other systems that focus on the way energy travels up the spine, the Celtic weave is portrayed as a series of figure eights, looping seven times up the torso. But this pattern can also be found throughout the body, over the top of the head, the face, the trunk, the legs, and the feet. These intertwining energies get ever smaller, eventually weaving their primordial pattern at the cellular level. In fact, a single strand of DNA, with its spiraling double helix, may be the prototype of the Celtic weave as an energy system. I am humbled as I look at the intelligence of the forces that weave the complex energies of a human body into a thriving unit. And this discussion gives only a glimpse. Within the Celtic weave are spirals, circles, and figure eights of bright light that connect and strengthen other energies and may also take form over a part of the body needing repair or extra support. I discuss these natural geometric miracles in a paper titled "Tibetan Energy Rings and the Celtic Weave," available as a free download from www.energymed.org/hbank/handouts/ tibetan_energy_rings.htm.

In addition to the Celtic Weave exercise are three other techniques for directly supporting the body's crossover patterns: the Homolateral Crossover, Rhythmic Eights, and the Suspension Bridge. Because the Celtic Weave exercise does "weave it all together," I present these first so that what the Celtic Weave exercise weaves is what you would want woven into your energy body. For each of these three techniques, I explain how you can know if you need them. Whether or not you *need* them, however, doing them will cause no harm and, like any good exercise, is generally beneficial. Then, with these other patterns in a good flow, doing the Celtic Weave exercise will tend to maintain that flow as it enhances the communication among all nine of your energy systems (see page 28).

Homolateral Crossover (time—about 4 minutes)

Your energies are running in a homolateral pattern if you are depressed, chronically exhausted, or ill, or if your recovery from even a mild condition seems slow. Your body does this to conserve its resources, temporarily sacrificing in its energy systems the crossover patterns that are necessary for optimal functioning. Too often, however, homolateral patterning becomes habitual. By some estimates, you have less than 50 percent capacity to heal when your energies are running in a homolateral pattern. So getting homolateral energies back into a crossover pattern, particularly during illness, depression, or simply when you have "lost your energy," is a fundamental energy medicine procedure.

Normally, the force of the Celtic weave keeps your energies crossing over, from the energies in the smallest cell to the way the right side of your brain controls the left side of your body and the left side of your brain controls the right side of your body. At the level of these large patterns, if your energies are going straight up and down the trunk of your body instead of crossing over, you are homolateral. While not uncommon, this is not good. You do not have your full vitality when your energies are not crossing over. You cannot think as clearly. You cannot heal as readily.

The Homolateral Crossover takes about four minutes. Most people experience immediate benefits in terms of their sense of vigor and capacity to think clearly. How long such benefits will last for you, however, depends on the state of the energies affecting your health and the energy patterns your body has established. If your energies have been in a homolateral pattern for a long time, or if your energies are depleted, you may need to do the Homolateral Crossover two or three times a day to form a crossover habit.

How do you know if the Homolateral Crossover will benefit you? One way is to simply try it and then see what you feel. There are also many subjective signs. As mentioned earlier, if doing the Cross Crawl is difficult for you, or if you cannot easily coordinate your opposite arms and legs, or if just starting to do the Cross Crawl confuses or exhausts you, your energies are probably running in a homolateral pattern. In that case, walking or marching in a natural cross-crawl manner is literally going against your own flow. Other important hints that the Homolateral Crossover might be helpful to you include depression, exhaustion, inability to get well, inability to benefit from exercises (include the Daily Energy Routine), or a lack of a response to treatments (conventional or alternative) that you have reason to believe should be effective.

There is also an energy test to determine whether your energies are homolateral, so this is an opportunity, if you have learned energy testing from the Appendix or through another source, to experiment with it. This will require a partner, and you can energy test each other. To do the energy test:

1. Draw a large *X* on one piece of paper and two parallel lines on another.
2. Look at the X. Have your partner energy test you.
3. Look at the parallel lines. Energy test again.

If your energies are crossing over properly, looking at the X will be in harmony with the crossover flow of your energies and you will test strong. Looking at the parallel lines will momentarily jar your energies, and you will test weak. But if your energies are homolateral, the opposite will occur. The X will weaken you, the parallel lines will strengthen you.

To get your energies to cross over again, rather than going against the homolateral energies, the Homolateral Crossover begins with movements that are in alignment with the way your energies are already flowing. It then guides your energies into a crossover pattern by doing a Cross Crawl. Going back and forth between these two movements stabilizes a crossover pattern. To do the Homolateral Crossover:

1. Begin by tapping or massaging the K-27 points (the first points of the Three Thumps, page 53) followed by a full-body stretch that "reaches for the stars."
2. Do a homolateral crawl, lifting the right arm with the right leg, and then

the left arm with the left leg. Do about 12 lifts (see Figure 2-12). This may be done standing, sitting, or lying down.

3. Then march in place, as when doing a Cross Crawl, lifting the right arm with the left leg and then the left arm with the right leg (see Figure 2-6, page 56). If you are sitting or lying, the movement can easily be modified to fit your position.

4. After about 12 lifts of the arms and legs in this cross-crawl pattern, stop and return to the homolateral pattern (lifting the same-side arms and legs) for about 12 lifts.

5. Then stop and return to a normal Cross Crawl (lifting opposite arms and legs) for about 12 lifts.

6. Do this sequence a total of three times.

7. Anchor it with an additional dozen normal Cross Crawls. End by again stimulating the K-27 points.

Figure 2-12
HOMOLATERAL CROSSOVER

Rhythmic Eights (time—about a minute)

Where the Homolateral Crossover addresses large patterns within the Celtic weave, doing Rhythmic Eights brings all of the energies, from those that surround you down to those of individual cells, into a crossover pattern. If you are having difficulty relaxing, are caught up in your daily troubles, or are just feeling the blahs, Rhythmic Eights are fun to do and are always a good energy boost. It adds a touch of joy to do them to music that makes you want to move.

1. With your hands hanging, sway your body, shifting your weight from one hip to the other, moving to music or moving as if you are moving to music.

Figure 2-13
RHYTHMIC EIGHTS

2. Let your arms sway with your body. You will notice that there is a natural figure-eight movement in both your arms and body as you sway from side to side. Let your arms move out farther so they are swinging wide.

3. Stretch your hands out in front of you and make a sideways figure eight with your arms. Go up and over on the right, circle down, then up and over on the left, circle down, and then again up and over to the right. Twist at the waist in the direction toward which you are reaching, allowing your body's rhythm to move with your arms. It is more like a free-form dance than a tightly patterned exercise (see Figure 2-13).

Several Rhythmic Eights will help get the left- and right-brain hemispheres into better communication. This technique is used in "Educational Kinesiology," an application of energy medicine that helps children with dyslexia and other learning handicaps. Several studies have shown that making rhythmic eights on the blackboard with chalk, while putting their whole body into the motion, helps children with learning disorders.[10]

Suspension Bridge (time—30 to 60 seconds)

Another way to get the energies in your body moving and crossing over is to elongate and stretch your spine in a crossover pattern. If your muscles are tight or your body doesn't have the flexibility you would like, this exercise can be rejuvenating.

1. Stand with your feet wider apart than the width of your shoulders.
2. Place your hands on your thighs above your bent knees and straighten your arms. Take several deep breaths. The position feels like sitting on an invisible chair.

3. With your head forward and your bottom back, adjust your feet so your knees are directly above your ankles and your back is straight. You are creating a kind of suspension bridge.

4. Slowly stretch one shoulder across and down toward the opposite knee (see Figure 2-14). Repeat with the other shoulder. It is a crossover exercise. You will feel the stretch across your back. You can do this stretch more than once.

5. Very slowly raise yourself with your arms hanging down, until you are in a standing position.

Figure 2-14
SUSPENSION BRIDGE

Fortifying Your Aura and Crossover Energies with the Celtic Weave (time—about a minute)

Your aura is a multilayered sphere of energy that emanates from your body and interacts with the atmosphere of the earth. It is itself a protective atmosphere that surrounds you, filtering out many of the energies you encounter, and drawing in others that you need. The health of your aura reflects the health of your body; the health of your body reflects the health of your aura.

When you feel happy, charismatic, and spirited, your aura may fill an entire room. When you are sad, despondent, and somber, your aura crashes in on you, forming an energetic shell that isolates you from the world. Sometimes if I feel closed in by too many intruding energies, I will very slowly place my open palms against the inside band of my auric field and push it away. If I go slowly enough, I can feel the pressure of the field as I push on it. Try it sometime when you are having difficulty claiming your space in the world, or are feeling sad, small, or squashed. Imagine yourself surrounded by an eggshell-like energy and exhale slowly as you push it away from you, beginning at about two inches from your body. It is easier to do than you might think. You may be able to feel an energetic force against your hands, and in any case, it will give you more breathing room, energetically and psychologically.

The Celtic Weave exercise keeps the energies within your body crossing over and also helps keep your aura strong and healthy. When these two energy systems are in good repair and working in harmony, your overall health is greatly supported. To do the Celtic Weave:

1. Stand tall, with your hands on your thighs. Throughout the exercise, breathe slowly and deeply, in through your nose and out through your mouth.

Figure 2-15
CELTIC WEAVE

2. Swing your arms around and bring them into a prayer position in front of your chest.

3. Rub your hands together, shake them off, face your palms toward each other, and see if you can feel the energy between them. Do not be concerned if you cannot feel it. Sensitivity will grow if you keep working with energy.

4. Rub your hands together again, shake them off, put them up next to your ears, about three inches out, and take a deep breath (see Figure 2-15a). Exhale.

5. With your next in-breath, bring your elbows together.

6. On the out-breath, cross your arms (see Figure 2-15b) and swing them out (see Figure 2-15c).

7. Cross them again in front of your body and swing them out again.

8. Do this again, but as you swing out, bend down and cross your arms over the top part of your legs (see Figure 2-15d).

9. Stay bent. Swing out again, in front of your ankles.

10. Bend your knees slightly. Turn your hands toward the front, scoop up that energy, stand, raise your arms, and pour that energy down the front, sides, and back of your body.

REPROGRAMMING DEEP-SEATED ENERGY PATTERNS IN BODY AND MIND

Energy runs in habits. Like flowing water that forms a riverbed, energy patterns that are repeated day after day tend to become entrenched. They are difficult to change even if they are no longer needed or useful. If one of your ancestors was born during a famine and meals were scarce and inadequate, her body would learn to turn protein that was optimally meant to build muscle into fat storage. When the famine ended, and food was plentiful, the energy systems that control metabolism would still be storing fat, now unnecessarily. This dynamic is true of the energies that govern your mind as well as those that govern your body. Your famine-born ancestor might be obsessed with food in ways her sibling, born four years later and after the end of the famine, might not be.

Fortunately, it is possible to change deep-seated energy habits that affect our bodies and minds. David, Gary Craig, and I wrote an entire book on this topic, *The Promise of Energy Psychology,* and we hear amazing stories about its effectiveness (see Gary's Web site, www.emofree.com). But I do not want to be glib about this topic. As Western medicine has slowly embraced the concept of the mind-body connection, many well-meaning health practitioners tell people to change their attitudes, beliefs, or self-concept as if it is like changing a pair of shoes. Too often, however, they don't give them the tools to do it, even though those tools exist. While I am a fan of the power of intention, simply willing yourself to change a belief or other mental pattern that has become deeply embedded in your energy system often has the effect of just giving you another reason to feel bad about yourself, beyond having recognized the self-defeating mental pattern in the first place. You can't think yourself out of a dark room, and you can't usually think yourself out of a major depression, the desire for a cigarette, or the longing for love from someone who is not available to give it. Shifting your energies, however, creates a new internal atmosphere that supports new ways of feeling and thinking.

I offer here three powerful techniques you can experiment with and integrate with the other methods presented throughout this book. The Blow-out/Zip-up/Hook-in, the Temporal Tap, and the Meridian Energy Tap each combine your focused intention with a way of shifting the energies that maintain the mental habit you wish to change.

Blow-out/Zip-up/Hook-in (time—one to two minutes)

When you are troubled, upset, or simply frustrated that you have not been able to change a pattern in your life, tune in to the feeling and:

1. Stand up straight. Put your arms out in front of you, bend your elbows slightly, make fists with the insides of your wrists facing up, and take a very full breath (see Figure 2-16).
2. Swing your arms behind you and up over your head. Hold here for a moment.
3. Reach way up, turn your fists so your fisted fingers are facing each other, and rush your arms down the front of your body as you emphatically open your hands. Let out your breath and your emotions with a "whooooosh" sound or any other sounds that come naturally.

4. Repeat several times. This will feel good. The last time, bring your arms down in a slow and controlled manner, blowing your breath out of your mouth as you go. Stand tall and breathe deeply.

5. In this moment of emptiness and release, formulate a simple statement about yourself that you wish were true. "I am calm and at peace." "I am feeling strong and confident." "I am a force to be reckoned with." Affirmations of this nature are in the first person, present tense, stated as if they are already true.

6. Zip the affirmation into body and mind by stating it slowly and deliberately while doing a modified Zip-up: Place your hands at your pubic bone, take a deep in-breath, and while stating your affirmation, move your hands, slowly and deliberately, straight up the center of your body to your lower lip. Do this three times.

Figure 2-16
BLOW-OUT

7. Lock it in with a Hook-up: Breathing in through your nose and out through your mouth, place the middle finger of one hand between your eyebrows above the bridge of your nose, the middle finger of the other hand in your navel, and gently press each finger into your skin (see page 62, Figure 2-11). Pull your fingers upward and hold for about 15 seconds.

Temporal Tap (time—about a minute)

The Temporal Tap was used for pain control in ancient China, but it is now being recognized as a surprisingly effective method for breaking harmful habits and simultaneously establishing desired ones. Tapping around the temporal bone—beginning at the temples and traveling around the back sides of the ears—makes the brain more receptive to learning while temporarily suspending other sensory input. It also relaxes triple warmer, the energy system that maintains physiological habits. This allows you to more easily slip in a new habit.

In the 1970s, the founder of Applied Kinesiology, George Goodheart, discovered that by tapping along the cranial suture line that starts between the temporal and sphenoidal bones, you can temporarily shift the mechanisms that filter sensory input. If you introduce a self-suggestion or spoken affirmation while you are tapping, your mind will be particularly receptive to it. The Temporal Tap takes advantage of the differences between the left and right hemispheres of the cerebral cortex. In most people, the brain's left hemisphere is more critical and skeptical than the right hemisphere. So statements made with a negative wording are in closer accord with the way the left hemisphere functions and are more likely to be assimilated. Such statements are tapped into the left side of the head. Likewise, because the right hemisphere is highly receptive to favorable input, statements that are worded positively are tapped into the right side. This pattern is reversed in some left-handed people, and you can energy test to see which works for you. If you stay strong while tapping a negative statement into the left side, stay with the instructions; if you become weak, swap the words "right" and "left" in the instructions.

Begin by identifying a habit, an attitude, an automatic emotional response, or a health condition you would like to change. Describe the change you would like to bring about in a single sentence, stated as an affirmation in present time, that is, as if the desired condition already exists. It does not have to be a truth yet, but rather a statement that you would like to be a truth in the future. For instance, you might say, "Under pressure, I stay calm and centered." It usually helps to write out your statement.

Then restate it, keeping the same meaning, but with a negative wording ("no," "not," "never," "don't," "won't," etc.). Thus the "calm and centered" statement may be worded negatively as "I no longer get stressed under pressure." Notice that the meaning is still positive, even with the negative wording. Another example: "I eat for health and fitness" can be negatively worded as "I don't eat from anxiety or compulsion." "My fingernails are growing long and healthy" can be negatively worded as "I no longer bite my nails."

1. Starting at the temple, tap the left side of your head from front to back with the three middle fingers of your left hand (see Figure 2-17). State the negatively worded version of your statement in rhythm as you tap. Tap hard enough to feel a firm contact and a bit of a bounce. Tap from the front to the back about five times, making your statement with each pass.

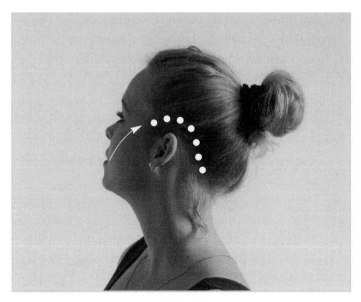

Figure 2-17
TEMPORAL TAP

2. Repeat the technique on the right side, tapping with your right hand, but this time using your statement in its positively worded form.

3. Repeat the procedure several times per day. The more you tap in the affirmation, the quicker and stronger the effect on your nervous system and your field of habit.

The Temporal Tap combines a variety of powerful elements, including repetition, autosuggestion, and neurological reprogramming. It affects not only the brain but also each meridian, so the message of your intention is carried to every system in your body. It is a disarmingly simple way to change many patterns that cannot be overcome by willpower alone.

The parallels between our physical and our emotional difficulties are often striking. Sometimes an organ will literally begin to behave in a way that reflects the way *you* behave. I'll use myself as an example. I sometimes have difficulty setting clear boundaries with people. Given the nature of my work, this can be an overwhelming problem. At one point some years ago, I became very vulnerable to infections. It was as if my thymus gland, which is responsible for protecting the body from infection, was modeling *its* behavior after mine. I began to tap into my left temple the statement, "My thymus gland no longer allows foreign invaders to enter my system," and into the right, "My thymus gland vigilantly maintains strong boundaries that keep out foreign invaders." Not only did my susceptibility to infections decrease, my ability to set clear boundaries improved as well. Part of the beauty of the Temporal Tap is that in funneling your intention down to two short statements, you can sometimes get to the nub of a complex and puzzling relationship among body, mind, and spirit.

The Temporal Tap is such an important tool for taking greater control of your life that I want to relate a number of stories where *it didn't work,* as these stories are instructive for getting it to work. I have found that when the Temporal Tap fails, it is often because of how the statements are worded. The words need to be in a language that is comfortable for you. Sometimes it just means translating the words into the simplest words possible. It is also important that the words be aligned with your values and congruent with your feelings, that you not be saying one thing while thinking another, and that you not be attempting to negate a primal need.

A woman who wanted to lose twelve pounds gained eighteen using this approach. On the left side, she tapped in the words, "I no longer hold on to extra weight." On the right side: "My set point is dropping to 134." These wordings seemed reasonable, but the tapping was having the reverse effect. I asked her to notice if her thoughts were following her words as she tapped, if her mind was wandering, or if any images were entering her awareness. It turned out that what came into her mind *every* time she tapped was, "Oh, hell, I've got a Slavic body, I'm always going to have a Slavic body, and I'm going to end up looking just like my [fat] Aunt Sophie." She was tapping this in five times every day. And it was working! The thought and image were far more primal self-suggestions than her carefully worded statements. She did, by the way, lose a good deal of weight after tapping into her left side, "I did not inherit Aunt Sophie's body" and "I inherited a body that can be thin and lithesome" into her right side.

A man whose job was eliminated after twenty-four years with a company was sent back to college for a year, at the firm's expense, to upgrade his skills. Five weeks into the first quarter, he had achieved Cs and Ds on the midterms of several of his courses, threatening his future reemployment. He was trying very hard, but he found himself unable to concentrate during any sustained period of study. Having heard that I had helped people with learning disabilities, he scheduled a session with me. I could not find a learning disability, but I did sense that after working for a quarter of a century in a people-oriented position, his habit field had no juice for inwardly oriented, concentrated study. I explained this to him, and he was highly enthusiastic after I taught him how to use the Temporal Tap for changing the habit field that was affecting his ability to study. But he came back a week later deeply discouraged. He had followed my instructions to the letter and devoted all the time required, but he could detect no change in his study habits.

He told me the statements he had been tapping in, and I watched him do the tapping. The words sounded good to me, but his energies weren't receiving them very well. As I explained how important it was that the words must sound right to him, I learned that he had come to the United States at age six and that English was his second language. After doing the Temporal Tap for another week, this time using his native language to make the same statements, he reported respectable improvements in his ability to read and concentrate.

A friend wanted to quit smoking after fighting a chronic cough for more than a year. I taught her the Temporal Tap. As she tapped in her statements—"I no longer smoke" on the left side and "I enjoy my cigarette-free existence" on the right—she started to smoke like crazy, and she felt miserable. Something about the tapping was making the habit stronger, and it was also making her extremely anxious. But since the tapping was having an undeniable effect, even if it was the opposite of the desired one, she knew it was potent and decided to figure out how to use it to her benefit. She sensed that her anxiety was a key to the problem. She figured out that saying "I no longer smoke" was triggering her anxiety, in part because smoking was her major method of relaxation. Playing her intuition, she found new words. She began to tap into the left side "Anxiety no longer causes me to smoke" and into the right side "I only smoke now for my highest health and pleasure." The anxiety dramatically receded, and she went down from a pack per day to about three cigarettes per week. She kept it at that level for several months, using her occasional cigarette as a mantra, an act of meditation. Her cough diminished. Eventually she quit smoking completely.

As with many energy medicine techniques, the Temporal Tap often requires more finesse than simply following a pat formula. To summarize some of the considerations to keep in mind, your Temporal Tap statements will be more effective if: (1) the wording of the statement tapped into the left side of your head contains a negative word and the statement tapped into the right side contains only positive words, (2) the statements are in harmony with the way you naturally talk and think, (3) as you say the statements, you keep your attention on the words and their meaning, and (4) the statements do not instruct you to do something that contradicts a core value or a more basic need.

I have seen the Temporal Tap be effective for an enormously wide range of problems. It has helped people stop smoking, drinking, overeating, and scratching compulsively. It has assisted them in building confidence, optimism, and self-esteem. It has aided them in stimulating their immune systems when fighting a serious illness, improving their metabolisms when trying to lose weight, and enhancing their coordination when trying to learn a new skill. It has been a factor in reducing tumors, clearing eczema, and lowering blood pressure.

Select a target behavior, emotion, or physiological condition that you want to change. Craft a negatively worded statement to tap around your left ear and a positively worded statement to tap around your right ear. You can test whether your statement is in harmony with your energies by tapping with one hand while making your statement and using the other hand as a friend energy tests you. Temporal Tap your statements four or five times a day for at least a week. With deeply entrenched habits, I have seen it take up to thirty days before the results became evident. Even if you are skeptical, judge for yourself based on the results you attain.

Meridian Energy Tap (time—two minutes)

Tapping on various sets of acupressure points while bringing to mind a psychological problem seems to rapidly shift the neurology that maintains that problem.[11] Using the same approach, improvements in physical problems are also increasingly being reported. While energy psychology, as it is called, can be easily learned and our book on the topic is readily available,[12] it is beyond the scope of this book to do it justice. But I present one simple technique that is very powerful and well worth knowing.

If you are able to stimulate a series of acupressure points distributed on different meridian pathways, you can pump your meridian system into action. Where pumping adrenaline through the body arouses the defense system, pumping energy through the meridians brings new vitality to body and mind, often bringing about unanticipated positive effects. It is always good for the immune system. Sometimes it corrects minor physical problems without ever having focused on them. It can also decrease anxiety and enhance your sense of well-being. It is an all-purpose intervention. And it is easy to do.

If you do it while mentally focusing on a problem, it has an added benefit. Focusing on a problem causes the body to go into whatever stress reactions are associated with that problem. If you simultaneously energize your meridian system, you can often neutralize that stress reaction, and in many cases prevent it from occurring the next time that same problem is encountered. So this is a good one to do when you are distraught, thinking about something that bothers you, or simply wanting a boost.

Figure 2-18
MERIDIAN
ENERGY TAP

1. Stretch your forehead to get blood flowing by placing your thumbs at your temples on the sides of your head, curling your fingers, and resting your fingertips on the middle of your forehead. Slowly, and with some pressure, pull your fingers apart so that you are stretching the skin. Breathe deeply.

2. Tap each of the following points for the length of one deep vigorous breath (see Figure 2-18). It is a firm brisk tap, but of course not so hard as to hurt or bruise yourself:

   a. The points on the inside edges of your eyebrows

   b. The points on the outsides of your eyes

c. Tap with all your fingers at your temples and continue tapping up, over, around the back, and to the bottoms of your ears. Do this three or four times.

d. The points under your eyes at the tops of your cheekbones

e. The point between your top lip and your nose

f. The point between your bottom lip and your chin

g. The three thumps (pages 53–55) in the order of K-27, thymus, and spleen

h. The points on the outsides of your legs between the knees and the hips (hang your arms, bend your middle fingers toward your legs, and start tapping)

This chapter has been a primer of basic energy medicine techniques that enhance the flow of the energies in your body. The following chapter is the first to present energy techniques that specifically address managing your hormones, focusing on those having to do with your immune system and with stress. The chapter begins, however, by examining the chemistry of hormones, their role in your health, and how medical politics influence the ways we understand them.

Chapter 3

Dancing with Your Hormones

HORMONES
More Unique Than Your Thumbprint!
More Troublesome Than Your Thighs!
More Potent Than Your Therapist!

Every woman I know wants to enjoy strength, vitality, and vigor.
She wants to maintain optimum health, radiant skin, and luxurious hair.
She also wants to stay physically active and vital, and still be able to think
clearly and remember facts and events. That's where hormones come in.

—SUSAN LARK, M.D.
Hormone Revolution

Hormones play our lives like instruments.

—D. LINDSEY BERKSON
Hormone Deception

Your brain is not your only source of intelligence. If you think of your *entire body* as a large brain imbued with incredible intelligence, you are closer to the truth. And your *hormones* are the chemical messengers that keep the whole system adapting to your world, moment by moment. Being able to influence your hormones with understanding and wisdom gives you a strong edge for adapting to a world that is radically different from the one in which our bodies evolved.

YOUR BODY'S SIGNALING SYSTEM

Energy medicine is a major key for bringing that understanding and wisdom to the functioning of your hormones. Your body is a latticework of subtle energy fields, all of which impact your physiology, and hormone production can be directly influenced *by moving the energies in your body*. This demonstrable fact allows you to have control over your health and mood in ways that other cultures have understood but ours has not. Hormones are messengers that orchestrate the activity of your organs, tissue, and cells. They govern every aspect of human functioning, from digestion to reproduction, from thought to action. They rush from cell to cell, providing millions of instructions every second. But their messages have to deliver the right information for everything in your body and mind to work in balance and harmony. If your hormonal system begins to malfunction, your life can suddenly plummet into illness, despair, and misery. If hormonal disturbances are corrected, your health and mood can soar.

The two most important ideas in this book are:

1. For almost every health condition a woman faces, hormonal imbalances are in the foreground or in the background, and
2. You can influence your hormones and improve your health, happiness, and vitality by using energy techniques that are accessible and easy to learn.

Female hormones remain, in many ways, a mystery to the medical community. The biochemistry of hormones has been mapped, but in the critical area of individual differences, Western medicine and science are stumped. The same medication for the same symptoms will help one woman magnificently, not help another at all, and for a third will produce side effects that are more harmful than the original problem. In addition, enormous contradictions among sophisticated research studies on the dangers of hormone replacement therapy, as well as on many other aspects of a woman's biochemistry, have become a tremendous embarrassment to the medical community and underscore the fact that one woman's cure is another's calamity.

ONE SIZE FITS ALL: NOT!

A strength of energy medicine is that its methods can be tailored to the unique energies and needs of each person. Routinely giving two people the same medication or other treatment simply because they have the same symptoms is medical madness. We are too different from one another genetically, temperamentally, and energetically to base health care decisions on broad generalizations or on research showing that a percentage of people improve with a specific treatment. You deserve better, and you saw in Chapter 2 how it is possible to assess the compatibility between your body's energies and a treatment you are considering. Individual differences must be taken into account in any well-considered health care decision.

I know this too well because my body does not follow the rules. Medical protocols that make most people better often make me worse. During the summer of 1974, I participated in an experiment conducted by Dr. Dorian Paskowitz, the resident surfer and physician at that time at Palomar College, just north of San Diego. A garden with a variety of vegetables and herbs had been planted on the campus and, for one month, the two dozen of us in the experiment ate only what we picked, raw and uncooked, fresh from the ground. Most of the other participants lost weight and reported that they were thinking more clearly and feeling energized. But though I was exercising regularly and vigorously several times per week and ate no more than the others, I gained 12 pounds and felt dreadful.

Dr. Paskowitz was stumped. He was trying to prove how healthy a natural diet of live foods is for everyone (his challenge to conventional wisdom was, "How can you be fully alive when you are eating dead foods?"), and I was not fitting his theory. Why? I was caught in his enthusiasm for a natural diet and expected the experiment to work. I didn't cheat and sneak bags of Fritos at night. I didn't appear to be genetically different from everyone else. What caused me to respond so differently?

Over the years I've come to realize that the riddle has a one-word answer: hormones. My hormones *really* don't follow the rules. I was once brought to a hospital, passed out, and the treatment I was given almost killed me. I also had the worst PMS of everyone I'd ever known, an outcome of low progesterone and high estrogen levels, yet all the remedies I could find at the health food store made me worse. No doctor was ever able to figure out my fluctuating thyroid. I didn't match the Western models about hormones on much of anything. In fact, it is in part because my own endocrine system is so screwy that I've had to learn enough about hormones to be able to write and teach about them. But in my teens, twenties, and thirties, when such information would have been invaluable, I didn't have access to it.

BRINGING AN ENERGY APPROACH
TO HORMONAL CHAOS

Although Telisha sometimes felt challenged by the demands of raising four small children, she loved being a mom. However, like clockwork, for one week out of each month, she hated the role and herself in it. For the seven days prior to the start of her period, she found herself screaming at her kids with little or no provocation, retreating into bed crying regardless of what the children needed, and generally living in a dark gloom of despair and debilitation.

Then, again like clockwork, her period would start, her mood would lift, her competence would return, and she would sail through the next three weeks, though under a cloud of guilt for what had occurred during the previous bout of PMS and dread of the next one.

Telisha scheduled a session with me, hoping for help with her monthly ordeal but not really knowing what to expect. The session happened to coincide with the start of her PMS time. I found that every system in her body was being compromised by energy disturbances. One of Telisha's major meridians was overenergized and several were underenergized. Because you can very quickly correct such imbalances, I was able to experiment with her by shifting the flow of the energies and asking her how she felt after each intervention. Her body relaxed as each pathway was balanced, and she felt an immediate change in her mood.

She went home with several exercises I taught her for maintaining the new balance. She did them faithfully until her period started, and the days that would normally have been torture for her were mercifully unusual in that she did not experience the dark gloom or mood swings that were characteristic of her PMS experience. She thought she was cured. As the next cycle descended, however, she forgot about the exercises and was taken by surprise into the depths of another PMS nightmare. She was soon on my doorstep. The treatment was straightforward. We used the same procedures we'd used in the first session, but with more emphasis on the importance of continuing to practice the methods at home every month. She came to appreciate that her monthly cycle is a natural rhythm. She also learned that when she was feeling good, not in PMS, she could easily apply energy methods that would make the coming of her period less extreme and disturbing. The difference this made in her life was enormous. Two years later, I was pleased to learn that she was teaching other women how to manage their PMS at our local community clinic.

YOUR HORMONES IN ACTION

If you think of each cell in your body as a theater with a thousand stages, hormones raise and lower the curtains. A hormone may be sent from a distant neighborhood to raise the curtain and start the action on a stage that boosts your immune system or on another that regulates your metabolism. When the job is finished, another hormone lowers the curtain and the chemical actors nap until that curtain rises again. A hormone is a molecule, produced by a cell, which tells other cells, tissues, and organs what to do. Each hormone gives a precise instruction for raising or lowering a specific curtain, and in coordination with millions of other hormones, may turn you on sexually, create a hot flash, or increase the size of your breasts in preparation for pregnancy.

It was once thought that the brain and nervous system somehow regulated many of the processes now understood to be governed by the secretion of hormones. Rather than a top-down communication system, we have with our hormones a lively community inside with trillions of potent and seemingly independent decision makers. Hormones operate from three levels of distance: (1) those produced within a cell to regulate activities in that cell or in nearby cells, (2) those that travel a very short distance through an established duct, and (3) those that are produced by the endocrine glands, which are then secreted into the bloodstream to control events in cells and organs that may be far away.[1] Energy fields are vital in broadcasting the signals that make this enormous community of decision makers work in harmony.

It is difficult to grasp how potent your hormones actually are. The estrogen that regulates your menstrual cycle, influences your sex drive, and prepares you for pregnancy—secreted continually during your thirty or so childbearing years—will, according to some estimates, have a total weight less than that of an olive. A tiny fraction of that, at the time of puberty, transforms a girl's body into that of a woman. We are speaking of a very potent drug.

Remove the body's ability to produce certain hormones and death or debilitating illness follows. Even minuscule imbalances can inflict a range of serious maladies. Describing the pervasive role of hormones and the glands that produce them, and writing in the 1950s when their place in human functioning was just beginning to be fully appreciated, physician J. D. Ratcliff observed: "Our amazing endocrine glands take some part in virtually everything we do. Lift an eyelid—it was hormones that made sure sugar was in the blood to provide muscle power. Cut a finger and hormones will be there to help control inflammation and wall off infections. Hormones are our common denominators with animal life. Some of the movie starlet's sex hormones are identical with those of a whale; some of a prizefighter's pituitary products match those of a mouse."[2]

Any plant or animal that has more than one cell produces hormones—that is how its cells communicate. While nearly every type of human organ and tissue secretes hormones, the most familiar hormones are those released by the endocrine glands, including the adrenals, hypothalamus, ovaries, pancreas, parathyroid, pineal, pituitary, testes, thyroid, and thymus. As a group, these glands weigh only a few ounces, yet they are so decisive in our functioning that Ratcliff noted that "this tiny amount of tissue acts as a kind of council of ministers for the body . . . the body's master chemists." The endocrine glands dispatch their hormone messengers directly into the bloodstream with clarity of purpose and precision in destination. The mission of a single hormone might serve in the stimulation or suppression of growth, the activation or inhibition of the immune system, the regulation of metabolism, the preparation for activity (e.g., fighting, fleeing, or mating), or the management of a phase-of-life change (e.g., puberty, caring for offspring, menopause).

Most people do not normally think of hormones as a force in shaping their brains. We think of our brain as a given of our inheritance, with its wonderful capacities and any limitations built in and relatively fixed. But daily experiences literally change the physical structure of the brain. Learning builds new neurons. Our brains are continually being reconfigured,[3] and our hormones, according to Mona Lisa Shulz, M.D., even serve to "mold how our right and left brains process feelings and thoughts." In her groundbreaking *The New Feminine Brain*,[4] Schulz, who is both a neuropsychiatrist and a medical intuitive, explains how our brains adapt to our world, so your brain is literally a different kind of brain from your grandmother's brain. And we are at the cusp of the evolution of a "new feminine brain" that takes advantage of our deep intuition and the unique wiring of the female cortex, with its many more connections than men have between the right- and left-brain hemispheres. Shulz shows how understanding your hormones can help you develop your brain to new possibilities. Meanwhile, understanding the energies that govern your hormones gives you a special advantage.

YOUR ENERGIES TELL YOUR HORMONES WHAT TO DO

Hormones, in infinitesimal quantities, are able to pull off epic achievements because they are catalysts that *excite* your body's energies (the word "hormone" is, in fact, derived from the Greek *hormon,* which means "I excite" or "set in motion"). Hormones and energy impact each other in a mutual feedback loop. Hormones tell your cells, tissues, organs, and their energies what to do; your body's energies tell your hormones what to do. In each of the subsequent chapters of this book, you will learn to direct your energies to better manage the hormones that shape your journey through life.

Energy medicine is the art and science of working with the relationships between energy and chemistry to promote health, vitality, and well-being. The important part of the formula is not *energy* or *chemistry.* It is both. While changes in your chemistry (such as those you experience in your monthly cycle) change your energy, shifts in your energy also change your chemistry. This has been evident time and again in my practice. Often simple energy interventions have changed a woman's measurable hormone counts, restoring balance. While I don't know of any formal studies, I have worked with many nurses who had easy access to laboratory tests. I often encouraged them to take a reading prior to a session and another right after the session. Changes brought about during a session appeared in their white blood counts, blood sugar levels, thyroid levels, and estrogen levels.

FABRICATING THE MYTH THAT NATURE'S WAY IS THE WRONG WAY

There has been a great deal of confusion about hormones within science and medicine. When combined with deeply accepted cultural myths that devalue some of the distinguishing qualities of the female body and psyche, a fog of misinformation prevents most women from fully understanding how their bodies work. If you are like me, so far from the norm that it is critical to know how to step in and manage your own hormones, this is a terrible handicap. But even if your hormones operate "normally," you can be enormously empowered in managing your moods, your health, and your vitality if you understand your hormones. So accurate information is essential.

A 1963 article by New York gynecologist Robert A. Wilson and his wife, Thelma, a nurse, began: "The unpalatable truth must be faced that all postmenopausal women are castrates." The article, published in the *Journal of the American Geriatrics Society*, was followed three years later by Dr. Wilson's *Feminine Forever*,[5] a best-selling book that was widely serialized and excerpted in magazines and newspapers across the country. Wilson argued that menopause was a "deficiency disease," much like diabetes. The estrogen depletion associated with menopause, according to Wilson, led to "a horror of living decay." The menopause in this rendition stole not only a woman's health, but also her femininity and her sexuality, leading to debilitating depression, or at least a "vapid cow-like feeling." Postmenopausal women, he said, "exist rather than live," seeing the world "through a gray veil" and moving through their days "as docile harmless creatures missing most of life's values."[6]

While this characterization would not have flown in any of the thousands of cultures throughout history that have venerated their female elders, America bought it. Wilson's solution was to insist that women take estrogen pills to replace what their deficient bodies could no longer produce, and he founded a private trust for promoting estrogen therapy (with $1.3 million funding from the pharmaceutical industry). Soon "hormone replacement therapy pills," as the *Boston Globe* would later report, "sat like pastel candies inside the medicine cabinets of middle-aged women from coast to coast. They were easy to use, widely available, and touted as the pharmaceutical embodiment of Ponce de León's fountain of youth, promising youthful vigor, sharper thinking, and softer skin for women confronting menopause and the years after."[7]

By 1993, the science backing this development included many studies published in reputable medical journals such as *The Lancet* and the *Journal of the American Medical Association* (*JAMA*). They suggested that hormone replacement therapy in menopausal and postmenopausal women enhanced memory and reduced the risk of breast cancer, genital cancer, heart disease, stroke, and bone loss. Yet many women began to discontinue HRT within months of starting it because of side effects such as depression, mood swings, bleeding, fatigue, headaches, increased blood pressure, chest pains, varicose veins, water retention, decreased bladder control, and weight gain. By 2002, the only benefit that had not been contradicted by larger and better-designed studies involved the reduction of bone loss. That same year, an NIH study of 16,608 women, published in *JAMA*, reported that taking HRT for five years increased the risk of breast cancer by 26 percent, of stroke by 41 percent, and of heart disease by 29 percent. A subsequent report from the same study found that HRT doubled the risk of dementia.

HOW COULD SCIENCE GET IT SO WRONG?

The *Boston Globe* article "Hormone Therapy's Rise and Fall: Science Lost Its Way, and Women Lost Out" suggests: "The stellar rise and spectacular fall of hormone replacement therapy ranks among the biggest medical mistakes in history, fueled by a combination of weak science, relentless hype, the herd mentality of doctors, and women's dawning redefinition of menopause from an inevitable 'change of life' to a manageable condition."

Yep, women lost out. Many of the early studies were based on old and sometimes unreliable medical records that often did not indicate the strength of the medication or even the exact medication. Another error in the research was the failure to track individual differences. The same symptoms might stem from a variety of causes. For instance, you may not be producing the estrogen your body needs, but your neighbor may have similar symptoms caused by excess estrogen. Giving both of you the same medication because the research shows that a percentage of people improve with a specific treatment is indeed madness.

Wilson's crusade to rescue women from what he called the "tragedy" of menopause was given a huge platform by the pharmaceutical industry. After all, if menopause were just considered a natural occurrence instead of a disease, hundreds of millions of women throughout the world would remain an untapped market for the pharmaceuticals used to treat this "disease." Wilson's doctrine that menopause was a disease needing treatment was the perfect message for the pharmaceutical industry to spread in order to open "the largest commercial market in the world."[8]

Menopause is every girl child's future. Leslie Kenton points out, in her classic book on menopause, *Passage to Power,* that drug manufacturers "spend a fortune each year in an attempt to convince governments, doctors, and the public that . . . using their products is the only choice available to women" for battling the dreaded disease known as menopause.[9]

IS SCIENCE GETTING IT RIGHT YET?
THE POLITICS OF MEDICINE

You may or may not care about the intricacies and politics of scientific research, but if you go to your doctor, describe your symptoms, and are given one of the more than 3 billion prescriptions that will be written in the United States this year, there is good reason to care. *You are betting your well-being, if not your life, on the accuracy of scientific research.* More than 100,000 people die each year from taking those pills *as instructed.* You cannot just stare at the pills to decide whether they will actually help your unique body with your unique symptoms, whether the recommended dosage is too weak or too strong, or what unintended effects they might produce. You rely on the research findings to have answered these essential questions *before* your doctor settles on the prescription.

Scientific research has been considered the pillar of truth in our culture, the knowledge you can count on. Whereas so much of the information that comes your way is trying to get you to buy or believe something according to someone else's agenda, science has the authority of objectivity. Or does it? Well, to rephrase the old adage, people who love sausage and trust scientific findings shouldn't watch either of them being made. This is not to say that the basic scientific method is not one of humanity's greatest achievements—it is! But the tail of politics and economics is now wagging the dog of health care research. With those three billion prescriptions per year, depending on where that sacred canine points, some fierce infighting is involved.

The Boston Women's Health Book Collective (of *Our Bodies, Ourselves* fame) has summarized the "quiet but radical transformation" that has occurred in medical research over the past three decades.[10] Before 1970, government sources funded the vast majority of clinical research. Today pharmaceutical companies fund three out of four of the clinical trials published in the most highly respected medical journals, such as *JAMA,* the *New England Journal of Medicine,* and *The Lancet.* About 80 percent of the research the drug companies were doing back in 1991 was at least being conducted in universities, where various types of reviews and balances supported the researchers' independence and objectivity. By 2000, two-thirds of this research was being conducted by for-profit research centers.

"So what!" you may think. Research is research. To be published in a reputable journal, it must abide by scientific standards. Well, think again. Scientists employed by commercial enterprises know a great deal about how to get the results their organization wants to see. A common strategy is that if a study does not appear to be producing the findings the drug company wants the world to know, the study is discontinued so the unfavorable findings do not have to be reported. Another deceptive practice is to test the safety of a medication that will be taken mostly by older people on younger healthy subjects who are not as likely to experience side effects and who are not as likely to be taking other medications that might negatively interact with the drug being tested. The Women's Collective summarizes: "By controlling the design of the research, the criteria by which patients are selected, the analysis of the data, and the selection of results to be published, drug companies shape medical knowledge."

Do these technical points make *that much* difference? Two separate investigations of just that question were published in 2003, one in *JAMA,* the other in the *British Medical Journal.* Guess what? When a study was funded by the company that sold the product being studied, it was between 3.6 and 4 times more likely to show positive results for the product than if the researchers did not have a stake in the outcome. A subsequent study in *JAMA* looked to see if this ignoble distortion was still the case if only the highest-quality journals were considered. Yes, and more so. In the most prestigious studies, the sponsor's product was 5.3 times more likely to show positive results than in studies of the same product where the researchers were objective. These biased studies, however, when they appear in the medical journal, are not labeled with neon signs that say "biased material" or "Warning, the Surgeon General has determined that what you are about to read is crap." They are, rather, among the primary sources with which your doctor makes decisions about what will help you and whether it will put you in harm's way. And it doesn't stop with corrupted research practices. The Woman's Collective goes on to show how the agencies we count on to protect the interests of health consumers are increasingly compromised:

- The guidelines that inform and direct clinical practices are formed by committees dominated by experts who may have active financial ties to the very companies that produce the drugs and other products under consideration.
- About 70 percent of the continuing medical-education courses that doctors attend are paid for by drug and other medical-product manufacturers.
- "Objective" medical journals rely on revenues from pharmaceutical advertising and from selling reprints of published articles to corporate sponsors, who then distribute them to physicians.
- More than half the funding for the division of the U.S. Food and Drug Administration (FDA) that approves new drugs and oversees drug safety comes from user fees paid by drug companies.

In brief, financial gain is getting to be the health industry's new "objectivity."

Have you filled a prescription lately and been shocked by the cost? Consider 3 billion such prescriptions each year in the United States. While the research doesn't tell us with confidence whether they work or whether they are safe, the revenues to keep them being written exceed your wildest dreams. Spending on drug marketing, according to congressional investigators, jumped from $11 billion in 1997 to nearly $30 billion in 2005. Profit margins among the leading pharmaceutical companies are typically three to four times higher than for other Fortune 500 industries.

As for natural treatments, drug companies can make a profit only on what they can patent, and treatments based on natural plants cannot be patented. Therefore, a huge market has, for instance, been developed for synthetic progesterone (progestin). While progestin is generally less effective and has many more side effects than natural progesterone, it can be patented. Since natural progesterone cannot be patented, it is not advertised or promoted by the medical industry. You, along with your doctor, are less apt to know much about it (more about progesterone in subsequent chapters), nor for that matter are you likely to hear about energy medicine from your doctor. After all, the drug companies have not yet figured out how to patent energy. The missing information, misinformation, and contradictory information is not limited to HRT, but extends to some of the most vital areas where Western medicine meets a woman's health challenges, from mammograms to osteoporosis to PMS. In brief, the information generated by the complex web of drugmakers, researchers, and government regulators does not necessarily have your best interests at heart.

Most of the doctors I know bemoan the fact that their profession is controlled by a web of social and economic forces that are unrelated to their patients' well-being. The point of this little diatribe through the health care propaganda industry is not really to point fingers. I must admit, however, that I don't mind suggesting that a teeny bit of institutional bias might be entering the system some of those 3 billion times this year that a doctor reaches for a prescription pad instead of suggesting a more natural remedy. Or that the 100,000 to 300,000 people who will die this year from taking their medications exactly as prescribed might have been better protected. But mainly, because this is a self-help book, I am emphasizing this parody of science to encourage you to remain vigilant. Do not let your own perceptions and well-considered judgments about your body and its needs simply dissolve in the opinions of authorities. No matter how contradictory the information or how smooth the propaganda, you can counter the disempowering impact of such messages. Becoming the chief scientist in the laboratory of your own body's energies allows you to assess possible remedies for your ills by observing the subtle effects they have on your energy system. These, in turn, predict the likely long-term impact the remedies will have on your body.

THE MYTHS THAT KEEP US TRAPPED

The medical myths that impact our health choices work in concert with another set of myths that impact us even more personally. We live in a culture where beauty and youth are given more status and resources than wisdom, compassion, and ability combined. We live in a culture where we are continually assaulted with (and compare ourselves with) media images of the "perfect" woman: tall, willowy, weighing at least 20 percent less than her height requires, rarely looking older than 25, and with all flaws on her skin airbrushed away. We live in a culture where an exemplary woman, in a failed sexual-harassment case, was denied partnership in a top-ten accounting firm "because she needed to learn to walk, talk, and dress more femininely."[11] We live in a culture where you know exactly what I am talking about.

When Tina first came into my office, her very life force seemed depleted. She was exhausted and depressed. In her early fifties, she had spent her life trying to please men, trying to be pretty, trying to do everything she was "supposed" to do in order to be loved, and now it felt too late. Three men had left her for younger women. She was very angry. She had "done it right," she had followed the rules, she had sacrificed her own needs for others, she had gone for her collagen treatments, and none of it had gotten her what she wanted. She wasn't pretty enough, she wasn't thin enough, and now she wasn't young enough. And none of this was ever going to change. When menopause hit, it seemed the last blow. She believed it would all be downhill from here. And she was left with the inescapable fact that she hadn't lived her life for herself. Her own instincts and life force had been subjugated for . . . for what? Now in my office she presented a list of health problems that included hypothyroidism, compromised adrenals, menopausal symptoms, chronic fatigue, terrible depression, and thoughts of suicide.

Her physical ailments were clearly tied to her emotional problems. I began by focusing directly on the energy imbalances that were associated with the physical symptoms. A lifetime of denying your legitimate needs can leave your energies badly scrambled. Tina's life force (specifically, the vital energy that flows through the kidney meridian) had been dangerously depleted. As we nurtured her life force so that it pumped more vigorously through her body, she began to meet the world with a new sense of empowerment. As the energies shifted, so did her consciousness. She came to recognize that rather than enforced isolation, this was the first time in her life when she was not taking care of others and had an opportunity to really take care of herself. This in itself was exhilarating. Beyond that, the kidney meridian governs fear. She found a new courage to be herself without fear of rejection, and she blossomed in an internal atmosphere that didn't long for or seek anyone else's approval.

Tina's story isn't unusual. Pretty by most conventional standards, much of her life force as a young woman still went into working toward an airbrushed image that is not attainable in real life. Today, in the United States, more money is spent on cosmetics than on education or social services. Each minute, 2,055 jars of skin-care products and 1,484 tubes of lipstick are sold.[12] Each week, tens of thousands of women voluntarily have their bodies mutilated in deference to an image of beauty that defies nature. Women pay dearly to look thinner, younger, or bustier, only to find out that while cosmetic surgery cannot usually implant genuinely higher self-esteem, it often creates unanticipated problems. And it reflects a sad but widespread failure in self-affirmation. A vital and lonely challenge for women as they mature, according to Susun Weed, is that "they must recast their own opinion of beauty so that it includes old women."[13] The impact on our self-concept of the cultural messages around beauty and age are so pervasive that it is hard to get our minds beyond the assumptions of the "beauty myth."[14] This affects not only how you feel about yourself and what you do to influence your appearance, it also affects the way you care for your health. This was tragically brought home to me by my experiences in Fiji. When I lived there in the mid-1970s, I had never met a happier people. The Fijian women were robust, voluptuous, and joyful. Television was not introduced to the islands until 1995. Within three years, 11 percent of Fijian girls, comparing themselves with the slender actresses and models that they saw showcased, were bulimic, regularly purging themselves.

Our culture's pathological standards regarding beauty and weight infest other cultures as they gain the technology to find out how we think. It is not a pretty picture. If we had a department of mental health that was given the same power and resources as our war department, we would as a society vigorously and unambiguously teach girls how to appreciate and value their bodies as the unique gifts of nature that they are. Instead, we have created powerful industries that forcefully teach girls to negatively compare themselves with the unattainable images that Hollywood and Madison Avenue have somehow decided are the ideal look. We hypnotize our men with those images as well, and everyone loses.

Christiane Northrup, M.D., reflects that the "devaluation" of real female bodies within our society, combined with the cultural legacy that "male is superior to female, young is superior to old," provided fertile soil for Robert Wilson's misogynistic campaign: "Our better-living-through-chemistry society was poised to help us control our unruly female physiology through birth control pills during our reproductive years and estrogen during menopause."[15]

As women, we bought the message from the medical industry that redemption from the horror of menopause was to be found in estrogen replacement, and we continued to buy it even with mounting evidence that for many women the solution would create far worse problems than the condition it was meant to address. Embedded in our acceptance of this message is a series of other cultural myths that affect us all. We have learned that our natural rhythms and cycles are to be fought and controlled rather than valued and honored.

HOW RITUAL AND KNOWLEDGE CAN REEMPOWER US

The epoch passage from little girl into woman is not celebrated in our culture. It is not commemorated in a special ceremony like a confirmation, a bat mitzvah, or a wedding. At a time in our lives when we are, really, still little girls, needing guidance and nurturance and assurance about what is happening to us and where we are going and what it all means, we get very little of any of that unless we have extraordinary parents. And even for very wise parents, it is not easy to pluck a ritual out of the air when your culture is so ambivalent and confused about the passage between being a little girl and becoming a woman.

I, however, was part of a changing era. Women were rising to collectively celebrate who we were, perhaps for the first time in thousands of years, since before the patriarchy became the dominant social force.[16] And we wanted rituals to celebrate our daughters. We were waking up to what it means to be a woman, and we wanted to pass these sweetest of revelations along to the next generation.

The Woman's Liberation Movement carried a consciousness that for some of us was like a flower blossoming. Latent dimensions of our power and creativity and potentials were being rediscovered in the most exciting ways. The pendulum was swinging away from blindly accepting the status of a woman as depending upon being a stay-at-home mom, subservient to her husband, getting dinner on the table in time, and trying not to make trouble. (I remember a bumper sticker that captured our dawning realization about our "place" in society: "Well-behaved women never make history.")

This shift included exuberant talk about how to ritualize a young girl's period. I personally was excited, wanting to truly commemorate this great passage and help my daughters know it was a wonderful event happening in their bodies. But we had no role models. We could only imagine how it might have been wonderful for ourselves if our moms and dads had given us a ritual. But such wisdom had not been passed down to us. Our memories, instead, held visions of all the girls in seventh grade being marched off to see *the* film on menstruation.

In mine, we walked in a straight line to the room where the film would be shown, past the boys, all pointing, laughing, and taunting us. Somehow they all knew. There was humiliation and even crying among some of the girls. Then we watched a stupid film that portrayed the narrowest of facts, and it left us all afraid.

All the teachers were embarrassed and didn't know how to talk about it. Mrs. McDonald was younger but wanted, nonetheless, to give some advice and wisdom from her 24 years: "Just remember this: it is completely up to you not to get pregnant. Boys can't help it, so you must never 'make out.' If you get pregnant, you will miss the prom and be sent away to have a baby alone." Menstruation was still given the name "the curse" by many mothers along with these same dire warnings that you could get pregnant. There was no pill yet.

The Woman's Liberation Movement was a massive energized effort to change all this and bring everything about our cycles out into the open. We were going to empower our little girls with the joy of their cycles. Even though our society was not ritualizing a young girl's first period, I certainly would.

Unfortunately, my husband stepped in first. This was before either girl had even been out on a date. He had decided that he was going to have "the talk" with them. I was immediately concerned. "Are you sure you know what to say, Ray?" Ray was 12 years older than me, from a generation preceding mine, and he was decidedly not into Women's Lib. He assured me that he knew exactly what to say. And he did. In fact, what he said had a lot to do with my deciding to divorce him before the girls became full-fledged teenagers. In their talk, he told them calmly and firmly, "If you ever get pregnant, don't come home." End of discussion.

When my turn came, however, I didn't quite create the example for all parents to aspire to, either. For instance, Dondi's first period came earlier than I expected. She was only 11 and the day sneaked up on me. In my inept hunger to make her first period beautiful, I lovingly held her and spoke of how she was moving into womanhood and what that meant. I didn't know what I did wrong, but after a horrified look came over her, she started crying. She whimpered, "I want to die!" I had so held the dream of doing this perfectly and beautifully for her that I couldn't bear her response. I slapped her, like you might in panic slap a sleepwalker heading toward a ledge. It was the first and only time I ever hit her. As we talked further to try to recover the mess I'd made of this precious moment, I learned that she didn't want her childhood gone. She was hearing what I was telling her in black and white. That morning she was a little girl who could play and have fun, and now she thought she no longer was and no longer could. Now she was supposed to step into the shoes of a grown woman. But she still felt like a little girl, and her mind didn't want to go where I was seeming to lead her.

We all need to develop rituals and practices that are based on a solid understanding of our own nature and place in life. And medicine needs to develop procedures that are based on a solid understanding of our bodies and their not-so-solid energies. So far, it has not. But *you* can. As Judith Duerk ponders:

> *How might it have been different for you, if, on your first menstrual day, your mother had given you a bouquet of flowers and taken you to lunch, and then the two of you had gone to meet your father at the jeweler, where your ears were pierced, and your father bought you your first pair of earrings, and then you went with a few of your friends and your mother's friends to get your first lip coloring, and then you went, for the very first time, to the Women's Lodge, to learn the wisdom of women? How might your life be different?*[17]

ENERGY MEDICINE FOR YOUR HORMONES

For every human body, the cascading chemical dance that determines health and well-being rests with the proper interplay of three major hormones: insulin, adrenaline, and cortisol. Because of the stresses, pollutants, and diet that are part of modern life, all three of these hormones tend to be oversecreted. Diana Schwarzbein, M.D., points out that well-informed changes in nutrition and lifestyle habits "bring the major hormones back *down* into balance," and by balancing a few essential hormones, "you will balance all your hormones."[18]

Three ways to promote the optimal functioning of insulin, adrenaline, and cortisol are proper diet and exercise, and enhancing your energy system through a practice such as the Five-Minute Daily Energy Routine (page 51). While energy medicine has much to offer if these three major hormones get out of balance, the primary focus of this book is on hormones that are specific to a woman's body and her journey into the miracles and mysteries of menstruation, fertility, pregnancy, and menopause. One energy system, however, triple warmer, governs the production of adrenaline and cortisol and, in conjunction with spleen meridian, influences the production of insulin. The management of triple warmer will be our focus for the remainder of this chapter.

Triple Warmer. Triple warmer is one of the body's most potent and least understood energy systems. While the hypothalamus is considered the master gland because it helps regulate breathing, heart rate, body temperature, blood pressure, and hormone production, it teams with triple warmer in carrying out all these functions. Triple warmer, among evolution's great success stories, is charged with keeping you alive in much the way your local police force and fire department are charged with keeping you safe. It governs three of the body's most extraordinary mechanisms:

1. The immune system
2. The emergency response to threat (fight, flee, or freeze)
3. The ability to form physiological and behavioral habits for managing stress or threat.

Triple warmer's energies move in two ways. They follow along the path of triple warmer meridian, and they also are able to jump from this pathway and instantly go to any spot in the body where triple warmer's help is needed. This natural hyperlinking ability is a property of the radiant circuits (see page 32). Triple warmer simultaneously functions as a meridian and as a radiant circuit, and it carries the supreme responsibility of keeping you alive regardless of how hostile circumstances might become.

When you think about the brilliant design of the immune system (able to recognize and ward off menacing invaders) and the fight-or-flight response (setting you into life-preserving action, no contemplation required), you begin to appreciate how the force that manages these survival modes must certainly be one of evolution's greatest achievements. However, the fly in the evolutionary ointment is that the immune system and the fight-or-flight response are masterful adaptations to a world that no longer exists. The strategies used by triple warmer were designed to help you survive in the brutal world of your ancestors.

Adrenal Fatigue: The Stress Syndrome of the Twenty-first Century. We live in what has been referred to as "the most agonizing century in history," the vast improvements in quality of life and opportunities for personal fulfillment notwithstanding. The stresses and information overload you face every day could not have been imagined by your great-grandparents when they were your age. The sense of one's world and one's place in it, which once remained stable for generation after generation, changes every decade for us, with our sense of security dwindling in the process. A child psychologist told me recently that the *average* child today carries more anxiety than did children who were psychiatric patients a few decades ago. Our unmanaged stress, according to one study, is a greater risk factor for cancer and heart disease than cigarette smoking.

Adrenal fatigue has been called "the stress syndrome of the twenty-first century."[19] Your adrenals are two walnut-size glands that sit on top of your kidneys. They produce or contribute to the production of some 150 vital hormones that influence every major physiological process in your body, including the release of two of the three "major" hormones, adrenaline and cortisol. Both of these hormones help the body cope with stress. Cortisol also fosters immune-system function.

Triple Warmer and the Adrenals. Triple warmer, which governs the adrenals and thus the production of adrenaline and cortisol, is always on alert for a foreign invader. It triggers both the fight-or-flight response in the face of external threat and the immune response when the body takes in unfamiliar substances in food, water, or air. It was a brilliant strategy when our food, water, and air were pure and when external threats were limited to attacks by predators or crises such as low food supplies. But today, your immune system has to sort through any of tens of thousands of artificial chemicals that appear in our food. Not recognizing them on the evolutionary list of what is familiar and safe, it has to decide whether to initiate a biologically costly immune response to fight them off. Triple warmer also has to distinguish between real threats and stresses of modern life that do not require a full-fledged fight-or-flight reaction. When your computer dies just before you hit Save on an inspired, middle-of-the-night outpouring of glorious words for your speech next week, that is certainly a stress, but it does not warrant the same biologically intensive response as when a saber-toothed tiger walked into your ancestors' cave.

The physiological and psychological cost of these false alarms is substantial, invisibly draining your life force. The physiological changes of the fight-or-flight response include: heart rate may double or triple; blood pressure increases as the coronary arteries dilate; respiratory rate increases; muscle tension increases; hormones such as adrenaline, noradrenaline, cortisol, oxytocin, and vasopressin are released into the bloodstream; hydrochloric acid is secreted into the stomach; glucose is released from the liver; basal metabolic rate increases; blood leaves the forebrain and digestive tract and moves into the muscles and limbs; pupils dilate, improving eyesight; systems not essential for fighting or escaping, such as the immune, digestive, and sexual systems, virtually shut down. That's a lot of biological capital to spend each time your computer crashes, a friend is upset with you, your husband is late getting home from work, or your daughter uses a swear word. But it is the same circuitry and physiological response for any of these relatively minor disturbances as it was when the saber-toothed tiger walked into the cave. That's what I mean by saying that your body evolved for a world that no longer exists.

In addition to the physiological effects of the stress-response syndrome there are also psychological consequences. The ability to maintain perspective diminishes significantly, as do other logical functions. The tendency is to rely on instinctive or habitual stress-induced behavioral patterns rather than forming a creative response to the situation. In addition, anger or rage tends to accompany and support the fight response, while fear or panic generally accompanies and supports the flight response. Hysteria, feeling overwhelmed, or numbness tend to result when the fight or the flight response is activated but then inhibited or otherwise not acted upon.

Beyond the immediate costs, if the stress response is continually activated, the physiological consequences can lead to disorders of the immune and autonomic nervous systems. These include susceptibility to infection, autoimmune diseases, chronic anxiety, chronic fatigue, and depression. The body's ability to produce adrenaline and cortisol, two of its master hormones, also becomes exhausted.

Retraining Triple Warmer. In short, each of us landed in a world that tends to send triple warmer into overdrive, and many of our health problems result from, or are at least exacerbated by, this single evolutionary twist. The good news is that you can literally retrain triple warmer to not be on overdrive. This is, in fact, one of the most important gifts energy medicine has to offer.

For instance, when a stress (whatever its nature) sends you into the fight-or-flight response, and up to 80 percent of the blood leaves your forebrain to go into your arms and chest to fight, or into your into legs to run, the whole cascade of ancient chemicals prepares you for action you probably will not take. A simple technique from energy medicine called the Stress Dissolver lets you interrupt the process whenever you feel stressed or threatened. Lightly place the pads of your fingers on your forehead

Figure 3-1
THE STRESS DISSOLVER

and your thumbs on your temples (see Figure 3-1) and hold firmly but with no pressure for a minute or two. An alternative version is to lightly place the palm of one hand on your forehead and the palm of the other hand on the back of your head, just above your neck. Both versions activate points, called neurovasculars, which impact circulation and bring the blood back to your head, shifting you out of the stress response. Breathe deeply as you do the Stress Dissolver. It is a physiological contradiction to stay in this position and to be in fight-or-flight mode for more than a couple of minutes, so the fight-or-flight response subsides.

Partnering with Triple Warmer in an Emergency. While I was writing this chapter, I had a powerful experience using this and related procedures for working with triple warmer. We had retreated to Baja to write. I was driving early one evening, alone, and about to turn into a market, when I came upon a gruesome scene. A bus had just hit a young woman and one of its back tires had stopped directly on her pelvis, hip, and upper legs. Mine was the first car to arrive. I pulled over, jumped out, and went directly over to her. The bus driver, who looked like he was in shock himself, said, "No, no," when I got to her and gestured for me not to touch her, fearing, I'm sure, that an untrained person might inadvertently cause further harm. I don't speak Spanish, but I heard the words "Yo soy doctor" (I am a doctor) tumble out of my mouth. Everyone who was on the bus or who was now coming onto the scene cleared a space for me to be with her. Time stopped; everything seemed surreal. The girl was unconscious. I saw that her life force was leaving her. Her aura had collapsed completely into her body and the energies were draining out at her feet. I focused on what could be done to keep her alive. The tire sitting on her became secondary. I was on the right side of her body and immediately began to hold triple warmer points that would stabilize her energies and take her out of shock. These points are on both sides of the body. I signaled to the nearest person, a Mexican boy probably about 13, to hold the same points on the left side of her body, and he promptly mimicked what I was doing, somehow maneuvering around the tire.

The Triple Warmer Smoothie (time—20 seconds)

Triple warmer energy starts at the tip of the ring finger, travels up to the neck, behind the ears, and ends at the temples. You can use the electromagnetic energies of your hands to readily calm overactive triple warmer energy by tracing part of this pathway backward. Here are the steps:

1. Place your fingers at your temples. Hold for one deep breath, again breathing in through your nose and out through your mouth (see Figure 3-2a).
2. On another deep in-breath, slowly slide your fingers up and around your ears, smoothing the skin while maintaining some pressure (see Figure 3-2b).
3. On the out-breath, slide your fingers down and behind your ears, press them down the sides of your neck, and hang them on your shoulders.
4. Push your fingers into your shoulders and then, when you are ready, firmly drag them over the tops of your shoulders, and smooth them to the middle of your chest, with one arm resting on top of the other (see Figure 3-2c). This is the heart chakra. It brings you home to yourself.
5. Hold here for several deep breaths.

a

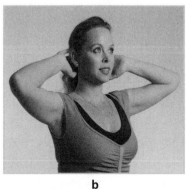
b

c

Figure 3-2
THE TRIPLE WARMER SMOOTHIE

The Triple Warmer Tap (time—about a minute or as needed)

This is a great "sedative" to use *while* you are feeling fear. It is simple, quick, and effective, sending impulses that turn off the fear or stress response.

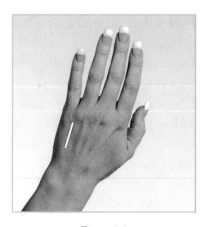

Figure 3-3
THE TRIPLE WARMER TAP

1. Place the hand to be tapped over the center of your chest.
2. With your other hand, tap it on the valley between the ring finger and little finger, above the knuckle (toward the wrist). Tap firmly, about 10 times, breathing deeply (see Figure 3–3).
3. Pause and take a deep breath.
4. Tap about 30 more times.
5. Repeat, switching hands.

We held these points until I felt the shock start to subside. While I was mostly on automatic pilot, I realize in retrospect that stabilizing her and starting to bring her out of shock was a necessary first step. Then other triple warmer points could be held. These would support her life force and get the energies that were draining out of her to remobilize and literally start to fight for her life. I suspect that we held the first set of points for about three minutes before it felt right to move to the second set of points. We just sat there on the ground with our fingers firmly touching these points. It felt like we were holding her life force in our hands. The boy was solid and steady in his task, but I noticed that he had tears in his eyes. Finally, I started to feel a power under my hands, as if her life force was coming back. As soon as this began to stabilize (I would guess that we held these points about three or four minutes), I was drawn to hold her neurovascular points (described on page 106). I showed the boy how to hold the end points of the heart meridian to give her heart all the support we could, and I went to the neurovascular points on her forehead. Besides interrupting the fight-or-flight response, which clearly was not adaptive in this situation, these points stabilize the circulatory system, helping balance the flow of blood throughout the body.

At about this time, an ambulance arrived with two paramedics. The bus driver immediately explained that I was a doctor. They approached to take over, but I felt it was essential to continue to stabilize her blood flow. I was able to indicate to them that they should let me continue, and they did. Somehow between the few Spanish words that were coming to me and whatever understanding they had of English, I was able to explain that the boy was holding points that stabilize the heart and I was holding points that could help with blood flow and internal bleeding. The paramedics turned their attention to how to get the bus off the woman. About five men had a brief discussion with the bus driver. Since the tire was directly over the woman's body, they needed to decide whether to roll forward or backward. I sensed that if they rolled forward they would damage more internal organs and indicated that they needed to roll backward. I also sensed that having the bus move over her again could put her back into shock and that the neurovascular points needed to further stabilize her first. At the moment that I could feel a pulse on her forehead, she opened her eyes and looked at me. Then she closed them again. I signaled to the paramedics and the driver that it was okay now to move the bus. I continued to hold the points on her forehead as the 32-passenger bus rolled back, off her pelvis, left hip, and leg. Almost instantly it seemed, she was on a stretcher, in the ambulance, and rushed off. The bus driver hugged me.

I was in a bit of shock myself. I went back to my car and cried. I then held my own neurovascular points, breathing deeply, until I felt I could drive. The next morning, I went back to the market to find out the hospital to which the woman was taken. When I got to the hospital, a woman who spoke broken English was able to identify the woman. She called a doctor, who came to greet me. The paramedics had apparently described the strange behavior of this blond gringo mystery woman. The doctor smiled, took my hands in his, and said, in good English, "You are the one who helped Maria at the accident. Something wonderful happened there. Somehow her blood coagulated and stopped her internal bleeding. Did it have something to do with what you did? It probably saved her life. What did you do?" I put my fingers on his forehead and said, "This is what I did." I've never seen appreciation turn more rapidly into bewilderment.

Ways to Partner with Triple Warmer. Hopefully, you will never need to use these techniques in such extreme circumstances, but you can use them whenever you feel fear that is out of proportion to the actual danger or whenever life triggers any other kind of stress response. By using one or more of the following

109

triple warmer techniques at such times, you are giving a direct command to triple warmer to not pump unnecessary adrenaline and cortisol into your system. You are partnering with it, training triple warmer to respond more appropriately to the stresses of modern life and literally helping it to evolve. If you use these procedures regularly, incorporating one or more of them into your Daily Energy Routine, for instance, you can help free yourself from the epidemic of triple warmer overwhelm facing all of us and the problems it causes with adrenaline, cortisol, and insulin, the hormones that regulate almost all the other hormones in your body. Two of the techniques have already been presented, the Meridian Energy Tap (page 80) and the Stress Dissolver (which is also described in greater detail on page 105). Here are four additional methods to consider for balancing triple warmer:

Figure 3-4
TRIPLE WARMER/
SPLEEN HUG

Triple Warmer/Spleen Hug
(time—about a minute)

This comforting position simultaneously calms triple warmer while energizing spleen meridian. Use it anytime you are upset or need comfort:

1. Wrap your left hand around your right arm, just above the elbow.
2. Wrap your right arm around the left side of your body, underneath your breast (see Figure 3-4).
3. Hold this position for at least three deep breaths.
4. Reverse sides.

Holding the Triple Warmer Sedating Points (time—6 minutes)

Holding your fingers on certain combinations of acupressure points can replenish depleted energies (these are called "strengthening points") or release jammed or stagnant energies ("sedating points"). The strengthening points strengthen meridians by *adding energy.* The sedating points strengthen meridians by *releasing excess energy.* Triple warmer is almost always overcharged in our stress-filled culture.

Sedating triple warmer not only allows it to function more effectively, but also keeps it from draining other meridians of their energies. Holding the sedating points accomplishes what the Triple Warmer Smoothie, Tap, and Hug accomplish, but with a deeper impact. Be sure to find a very comfortable position while holding the points. Holding acupressure points to sedate or strengthen your meridians is among the most time-consuming procedures presented in this book. But I love the results I get, as do my students and clients. And this is one procedure where you don't have to be mindful. You can hold the points while watching TV, talking to a friend, or relaxing in the tub. To sedate triple warmer meridian (the numbers in the meridian-strengthening and -sedating instructions throughout the book, as in "Stomach 36," refer to the name of the acupressure points being held):

1. Place the middle finger of one hand on Stomach 36 and simultaneously place the middle finger of your other hand on Triple Warmer 10 (see Figure 3-5a), on either side of your body.
2. Hold for about two minutes.
3. Repeat on the other side.
4. Then place the finger of one hand on Bladder 66 and the finger of the other hand on Triple Warmer 2 (see Figure 3-5b) and hold for about a minute.
5. Repeat on the other side.

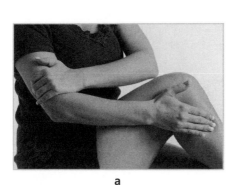
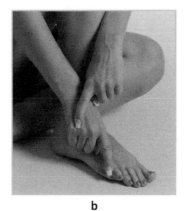

a   b

Figure 3-5
TRIPLE WARMER SEDATING POINTS

Energy medicine can help you understand your body in new ways and empower you with natural and highly effective methods to care for yourself. After examining some of the culture's distorted and insidious medical myths, this chapter has presented several basic methods for managing your body's master hormones, particularly in relationship to the stresses of daily life. In the following chapters you will learn how to use energy medicine methods with a variety of issues that specifically impact a woman's body, such as PMS, sexuality, fertility, pregnancy, and menopause. And you will be invited to learn how to combine energy interventions with energy tests to help you and your doctor determine the kinds and amounts of natural hormones or other supplements you may need. An energy approach does not necessarily replace the best of what medical science has to offer. But it can refine it, and it can go beyond mere chemical interventions. With knowledge about your body's energies, you can responsibly insist on being your doctor's collaborator in your treatment and, together, you will both be more effective.

Chapter 4

Reclaiming the Wisdom of
Your Menstrual Time

*We can reclaim the wisdom of the menstrual cycle
by tuning in to our cyclic nature and celebrating
it as a source of our female power.*

—Christiane Northrup, M.D.
Women's Bodies, Women's Wisdom

Premenstrual syndrome (PMS) affects sixty percent of the women in the United States. It is a medically diagnosable disorder and, with extreme PMS, a woman's very sense of sanity may feel assaulted. Beyond the physical symptoms, it is for many women no less than a ferocious wrestling match with forces from deep within. It is as if the soul is trying to burst out at the seams while the body's disorienting biochemical fluctuations are squashing it down. Our culture has not been particularly adept in preparing us for this challenge. Working with your body's energies not only may help you feel better, immediately and palpably, but also may shift your body's chemistry so that your journey through your monthly cycle is an adventure of the female soul rather than a fight with the female body.

After discussing the chemical, emotional, and energetic dynamics of a woman's cycle, this chapter coaches you on measures you can take throughout the month that will make your premenstrual time less troubling. It then presents energy techniques to do at the first signs of the onset of PMS and outlines a Premenstrual Module you can add to your Daily Energy Routine for bringing your body into better harmony with the chemical changes that occur prior to your period. It also offers specific energy techniques that can help with the physical discomfort and emotional challenges that may accompany PMS. Finally, it describes ways to energy test for hormonal and other supplements that may augment the energy approach.

IS PREMENSTRUAL SYNDROME A SYNDROME?

Hysteria and *uterus* have the same derivation (*hystera* means uterus in ancient Greek). Wide variations in the moods of many women, moods that correspond with their monthly cycle, have led to the use of terms like hysterical (or worse) to characterize some of the more extreme behavior resulting from monthly hormonal imbalances. Ranging from anxious to depressed, aggressive to weepy, and reactive to numb, these mood swings are described as part of premenstrual syndrome (PMS). Along with severe shifts in mood that are independent of events, conspicuous physiological changes may also accompany the time prior to menstruation, including bloating, swelling, fatigue, constipation, and skin changes.

At the same time, our attunement to the ebb and flow of natural cycles is a deep source of a woman's power. But when the medical establishment bestows a name like PMS on a natural process, that process is set on a slippery slope toward becoming pathologized and something to treat. We begin to move away from self-acceptance, away from understanding nature's plan, away from time-honored wisdom. Here we see the medical industry reshaping reality. First of all, the term "premenstrual syndrome" has a unique meaning for every woman. Some never experience it. For others, the symptoms last much of the month. For some, it seems related to lower levels of progesterone, in which case supplemental progesterone greatly reduces the most unpleasant features of their monthly cycle. For others, adding progesterone doesn't touch the misery they endure each month. PMS is a single term for many hormonal landscapes.

For all who experience it, however, PMS potentially holds a positive, creative, rehabilitative function. Like the sacred plants used in shamanic healing ceremonies to open a doorway to the world of spirit, PMS changes the body's chemistry to clear a path for consciously journeying into the domain of the soul. It unleashes the profound power of your inner regions, often in ways that women find overwhelming, bringing them face-to-face with great forces that usually reside in the netherworld. Viewing PMS as only a medical condition misses half the fun. Tribal cultures knew that the premenstrual days were a sacred time, a time to honor and respect, a time to sanctify in customs such as the moon hut.

Anyone who knows me, however, knows that I am not one to romanticize or become overly sentimental about the spiritual potentials of PMS. I know it in its darkness as well as its light, and in its darkness, PMS is a doozie.

YOUR MONTHLY CYCLE:
A PLAY OF TWO HORMONES

The seemingly tangled biochemistry of a woman's monthly cycle is actually among nature's most remarkable designs. While you are going about your daily business of eating, sleeping, working, shopping, playing, loving, and parenting, your body is busy choreographing a most amazing and time-honored pageant. One of the two million or so immature eggs stationed in your ovaries when you were born is selected as Queen-apparent and parades down the runway of the fallopian tubes toward the reproductive palace that is your uterus. If you have made the choices of nature's urging, she will be met by millions of suitors, choose one who has proven himself a most robust and healthy fellow, and they will join to grow into the newest achievement in humanity's mission to perpetuate itself.

The Queen-apparent has only about a day to find her suitor. If not, she is recycled (perhaps it was her hairdo), and preparations for a new Queen-apparent are set in motion. Nature seems harsh. But she wants only the best for you. And guess what? The complex stage directions for the pageant are coded largely in the release of two chemical messengers—yup, the hormones estrogen and progesterone.

However, other chemical messengers are also involved. In fact, the whole cycle begins when your hypothalamus tells your pituitary gland to release into your bloodstream follicle-stimulating hormone (FSH) and other hormones related to ovulation. Follicles are tiny sacs within your ovaries that house immature eggs. The hormones sent from your pituitary gland prompt the follicle holding the egg that is most ready to mature (the winner of the Queen-for-a-day competition) to get on with it. Showtime!

As the Queen-apparent puffs out her chest and steps toward the mike, the follicle splits, and she is suddenly free. Meanwhile, the blood supply to the ovaries increases and its ligaments contract, pulling the ovaries closer to the fallopian tubes, paving the way for the Queen-apparent's journey down the runway. The follicle's job, however, is not done. Not content to be an idle empty nest, it continues to support its offspring by rapidly taking on new cells so it becomes the ovaries' largest inhabitant and, now called the corpus luteum, begins to produce the estrogen and progesterone that orchestrate the pageant.

Estrogen directs the body to prepare a nursery for the much-anticipated child of the blessed excursion down the runway. The uterine lining is thickened so the nursery can contain the rapid growth of new life. Acting in tandem, progesterone causes glands within the uterus to produce secretions that stabilize the surface of its lining (the endometrium). Meanwhile, estrogen also gives the guys a hand by changing the consistency of the cervical mucus, making the juicy atmosphere more hospitable. "Fertile" cervical mucus draws the sperm into the fallopian tubes (where the festivities will take place) and helps them survive as they await the appearance of the lady of the hour.

But if, after all that, by the end of the cycle none of the guys have come through, or if they never showed up, the corpus luteum realizes that its chances for becoming a grandparent have passed, and it stops contributing its estrogen and progesterone to the family trust. Without these, blood vessels in the uterine wall contract and spasm, the uterine lining is shed, and the menstrual blood cleans all, initiating the start of a new cycle.

Not surprisingly, this internal royal life-and-death drama can have some effect on your mood. Even if you are oblivious to the plight of the Queen and her suitors, the hormones that manage the pageant may be harder to ignore. For some women the hormonal changes over a month are extreme; for others they are relatively minimal. Some women are also neurologically wired to have strong reactions to the slightest hormonal changes; others hardly notice massive fluctuations. The changes in estrogen and progesterone also trigger neurotransmitters, most specifically serotonin and gamma-aminobutyric acid (GABA), which can impact mood dramatically. If hormonal levels rise and fall too quickly, the adrenal glands become burdened and begin to secrete abnormal amounts of the stress hormone cortisol. This strains the entire endocrine system, making it ever more difficult to achieve hormonal balance as estrogen and progesterone are telling cells to multiply. Then, if the egg has not been fertilized and there is no baby for the uterus to carry, enzymes liquefy the uterine lining, the flow of blood begins, and the month's surplus uterine lining is shed. The entire body is involved. Even the bones are receptors for estrogen and progesterone. Is it possible to manage the ever-changing ingredients of this divinely inspired biochemical soup? I will share my journey.

MONTHLY VISITS TO HORMONAL HELL

In reflective moments, I like to believe that my many health challenges were a necessary part of my training as a healer. The concept of the "wounded healer"—that terrible illnesses and woundings establish a healer's expertise—resonates with me. Multiple sclerosis, tuberculosis, heart problems, a malignancy, and asthma have all been personal teachers for me. But beyond any of these, few people have been better prepared by their own life's journey to write a book on working with hormonal challenges.

My menses began when I was ten. Because I was so young, my mother had not yet had "the talk" with me. My panties were red with blood. I was dumbfounded. I thought I must have done something terrible, but I did not have a clue what it was. It was so disturbing that I decided to hide my panties in the attic so nobody would see. Mostly, I didn't want Mama to be scared. But with a nose for such deception, Mama quickly found them. When she saw me next, I received a knowing and compassionate, "Oh, Donna!" Oh, Mama, bless you!

Every month from that point onward, two days before the start of my period, terrible cramps, backache, and groin pain were accompanied by severe shifts in my perceptions and emotional reactions. The day before my period, the pain and pressure in my abdomen, groin, and at the top of my legs was so intense that I could hardly stand.

The experience is intensely biological. Women who suffer extreme PMS are, in fact, experiencing something akin to drug withdrawal, with a sharp decline in the body's natural opiates in the days prior to menstruation. It felt like I was being pulled low to the ground, toward the earth, and deep into myself. It was beyond choice. My legs felt heavy. The chemistry that fueled my usual exuberance completely waned. Energies from others and from the world now hurt my body as well as my psyche.

Then there was a day each month, just before my period started, when I was so awake that I would stay up all night, seeming to accomplish everything. The house would be made spotless, letters would be answered, papers filed, clutter organized. Nothing could tire me. I was in full joy and energy. Then, right before my period, I felt full of lead, pulled down, compelled to rest. This would actually sometimes knock me out, and I would go into long slumbers. They always felt to me like an ancient hibernating response. This was a time that my body needed to do major internal work, making a thousand adjustments as it readied itself to discard another month's preparations for carrying a baby and enter a new cycle.

David describes a day in my PMS life if I don't get to be moon-hut internal: "It is like I went to bed with Mrs. Jekyll and woke up with Mrs. Hyde. It is like the happiest, most joyful person I know becomes a ranting, tormented soul, sure she is responsible for healing the pain of every pitiful soul who passes her way. Any missteps of the past month are put under a microscope, magnified a dozen times, and dwelled upon endlessly, though it is a very democratic microscope—it is equally sharp with my missteps as it is with hers. Words meant to comfort are heard as evidence that I will never be able to understand her. Her misery is extreme, and my usual first-response tools—reason and repair—make things even worse. Was there nothing in my training as a psychologist, I would ask myself, that prepared me for each new installment that, like Groundhog Day, presented itself like clockwork, giving me yet another slim chance of getting it right?" I would like to add, David, that I was far more tormented than you were by not getting it right!

The role of hormones, though I didn't yet think in those terms, became patently clear during my first pregnancy. No monthly descent into physical and emotional agony! Also, for the first time in my life, I didn't have to watch my weight. It was as if my metabolism allowed me to process food like other people. Before, I could fast for days and not lose a pound, and it almost seemed that I could gain weight by merely breathing the aromas from the kitchen. Now I could eat whatever I wanted and, except for supporting the growing baby, be losing weight. I gained zero pounds during that pregnancy! And I ate and felt better than ever before. My body and my spirit loved being pregnant. And after my daughter Tanya was born totally healthy, I weighed 20 pounds less than I did when I conceived her.

I now understand that a woman produces a great deal of progesterone during pregnancy, which is necessary for building up the wall around the uterus and preparing the womb to carry a baby. Progesterone would later turn out to be my drug of choice. The elevated progesterone levels were, in part, why I loved being pregnant. I also loved being a mom. My progesterone production, however, went back to its former level, and PMS again became my private terror. After Tanya was born, my hormone-induced cramps and cyclic swings uncharitably returned, along with distortions of reality that could be dangerous. Our apartment was upstairs. Sometimes it would seem perfectly logical in my happy reveries as a young mother that if I dropped Tanya, sending her and her basket over the railing, she would gently float down to the bottom and I could then walk down the stairs, pick her up still comfortably snuggled in her basket, and continue into our day. Thankfully, a small voice from deep within was able to persuade me that this might not be a good idea. But I am not exaggerating to say that I had to fight the impulse, and I wonder how many mothers sit in jail for not having been able to sort such reveries from reality during the postpartum period. It had nothing to do with wanting to harm her or with postpartum blues, but rather a temporary inability to distinguish fantasy from reality.

When I was pregnant again with my second daughter, Dondi, I was equally happy and joyous, full of vigor, cheer, and good spirits. No cycles. No distortions. No pain. And after the birth, again the magic faded as abruptly as it had begun. It took some self-restraint not to have 15 or 20 children so that I could live as much of my life as possible enjoying the delights of pregnancy.

As the years went on, my sensitivities in the week before my period became even more extreme. As PMS approached each month, I would begin to feel trepidation. What was I going to do if I had a client scheduled or had to teach a class? I am a very positive person with lots of energy but, when PMS came, the bottom dropped out chemically, energetically, and emotionally. At this time of the month, other people's energies could fry my circuits. Especially irritating were positive, outgoing people, the kind everybody loves to be around. It was excruciating to feel my compulsion to respond to other people's coaxing or needs as my body and soul were pulling me inward. Yet I still tried to connect with them.

I never knew anyone who had PMS as bad as I did. Women who don't go through severe PMS often cannot even identify with my PMS experiences. Once, when one of my closest friends was furious at me for "losing it" during a highly inconvenient moment, she shouted, "Shouldn't you be held responsible for what you did?" I considered this for a moment and then replied with the certainty of fresh and profound insight, "No, I absolutely am not accountable!" Nothing even gave me a clue that I might be about to blow up. I might be perfectly normal, even happy and serene in one moment, and then something would trigger me, and I would feel nuts, finding myself screaming or in tears. I could be shielded from those triggers when I was able to take "moon hut" time. But with all the demands of parenting and career, that was infrequent, so I was often vulnerable.

Mine was certainly not a politically correct PMS. It was bad news for women to hear about severe PMS. Even my female gynecologist wanted me to conceal my experience, saying "Donna, I don't think I'd speak of this out in public and let people know just how bad it can be for a woman. After all, we want to get a woman into the presidency." Women were coming into their power at that time, and it did not seem that a person who had this condition could be head of her own affairs, much less of her country. They didn't want women showcasing their monthly impaired judgment and deep psychological struggles. They felt PMS gave women a bad rap. There was already a glass ceiling, and they didn't want anything to lower it. I, on the other hand, felt that these were the very qualities that the country needed—leaders who went within once a month and wrestled with their souls.

Nonetheless, I was terrified as PMS approached each month, and so were those who loved me. But not my clients. They were thrilled. People in my hometown would mark on their calendars when I was going to be premenstrual and try to schedule sessions at that time. Personal boundaries, which have not been my strong suit under any circumstances, were virtually nonexistent when I had PMS. It was impossible for me *not* to feel what was going on with anyone in my presence or not to offer every ounce of my life force. I knew their pains, the dynamics of their illnesses, and their personal sorrows as if they were mine, and my empathy and desire to help were far beyond reasonable. But what had people fighting to be given appointments during my premenstrual week was that I was extremely psychic during that time. The information I would receive about them was uncannily accurate. Having no boundaries does have its rewards.

But for me, working during this time was terrible. It threw my hormones off even further. At exactly the time everything in my being needed to turn inward, to retreat from the world, my sensitivities became so acute that I suffered for others to the degree that I would find a way to meet with them no matter what the personal cost or how late the hour. And there was no shortage of clients. Yet it was precisely at this time that I was not meant to be out in the world healing anyone. It was time to care for myself, to drop in deep, and see what I wasn't seeing. "Healer, heal thyself," the concept that had gotten me through so much in my life, got blanked from my screen while I was premenstrual.

THE UPSIDE OF THE DOWNSIDE

But now, a surprising revelation. I loved PMS! Not while it was happening, of course. But I loved it when it was over. And I love it now when I look back. Beyond the many horrendous parts of PMS, if I listened to my soul and my body, I found an upside to the downside. I cherish that I had such an uncontrollable force hit my body each month that humbled me and put me on a monthly spiritual path. My soul beckoned to me. This was not my persona. It was not a trick of my psyche or a chemical imbalance. I was wrestling with the deepest forces of my being. My truest nature was peering out through the cracks of my hormonally altered reality. This was a great gift, even if presented in frightening wrappings.

My body's insistence that I go deep within was not easily denied. It felt painful and wrong to come up from that place to interact with people. Positive energy felt negative. But the deep internal energy that drew me also felt scary as I began the descent. I felt anxiety and fear about all the things I couldn't control, most especially my own emotions. It felt as if I was falling down an elevator shaft, and I didn't know how far I was going to go.

But there was no escape. The wisdom of the feminine soul doesn't allow it. Difficult truths I'd been able to ignore or deny throughout the month came up large in my heart, often exaggerated into negatively tinged caricatures of themselves. The polarity from being external and out in the world to being internal and deep within myself was paralleled with a swing from light to dark. From this immersion into the dark night of the soul, I came back transformed. Hellish as it is, at least I didn't have to pay thousands of dollars to go on a vision quest. And if I had embraced the process—left worldly time and entered moon-hut time—I came back transformed for the better. As I began my ascent back up each month, I knew I was changed, touched by profound truths and a larger perspective. The journey was like meditating, but on steroids. I have always been a bit ambivalent about disciplined meditation practices since my body plummeted me into the depths this way. While recognizing the profound places meditation can take people, it seemed a genteel substitute, something perhaps for men or for women who didn't have the privilege of extreme PMS.

When I got to the other side of my period, I was always deeply grateful for it because I understood things about myself and my life that I don't believe I could have otherwise. Beyond its role in renewing our monthly biological ovulation cycle, we *need* PMS for its role in renewing our contact with the deep forces at the very heart of life. However, in our culture, which doesn't support women being in their own rhythm at that time of the month, we don't get to follow our soul's calling. Because we try to be out in the world and "normal," our PMS intensifies and sometimes becomes debilitating.

A woman who can fully appreciate her cycle, and the depths it can lead her to, develops wisdom that is emotionally rich and intuitive. She becomes more sensitive to her own intimate rhythms and the greater rhythms of life. She learns her body and its energetic ways as well as its hormonal ways. She is forced to surrender every month, and so she learns to embrace the quality of surrendering. The best part for me was how my soul *insisted* that I be in conscious contact with it. I could not ignore it, and I learned to surrender to its beckoning. The wisdom of my inner impulse was further supported when David got with the program by building me a "moon hut" (does it count if it has propane gas, a top-of-the-line Simmons bed, and a stereo system?) right in our backyard.

Despite the ways PMS yanked me off center, it was also my truth serum. Anything I had been able to deny or repress or not see would suddenly appear large on the screen of my inner life, with sharp tones and vivid emotions. Feelings that had been stuffed, truths that had been disregarded, and injustices that had been ignored throughout the month were in my face. Brutally in my face. I could not put up with a lie or arrogance or someone devaluing another. I had no censor in me. No tolerance for the nonsense we all accept as part of daily life. No way to hold back feelings that discretion might have kept restrained. I could not go along with anyone else's agenda. I often did not even know if my marriage would survive each month when I was premenstrual. PMS is nature's design for making you reexamine your life intensively for a few days every month, and few institutions can withstand such scrutiny (self-defense, by the way, was the *real* reason David built my moon hut!).

An elemental life force is always at work in a woman's body. While we are fertile, we enter deeply into this space every month. The cycle is both intimately personal and universal. The physical journey through the days before and during menstruation is miraculous, but PMS is also the passage through a doorway to a holy place. It is a sacred, intimate journey with the self and the soul.

During my PMS time, my dreams were often lucid and powerful. I sometimes couldn't tell in the daytime whether I was daydreaming or having psychic visions. In terms of yin and yang, it was extreme yin, extreme feminine. When I got to the other side of my period, I knew I was fortunate among women to know this extreme of being woman. While it seems on the surface that women who do not experience PMS are very lucky, I have secretly felt that missing out on the sacred journey of the worst kind of PMS is a great loss. Evolution seemed to speed up for me during premenstrual time, and my soul gave me guidance in a way that just wasn't possible in my normal busy days. The voice of my soul would call, and I *had to* answer. PMS is a path of power and wisdom but, in a world without "moon time," it can also be a path of distress and despair.

SEARCHING FOR THE MAGIC PILL

During clearheaded, upbeat times of the month, I began to experiment with ways that I might help myself during the challenging times. Doctors didn't have remedies for PMS beyond drugs designed to help with headaches and cramps. These did not touch my symptoms. I went to health food stores in search of relief. They offered several different herbs that supposedly helped with PMS. These consistently made my symptoms worse. I later learned that the active ingredient in all of these was natural estrogen. The popular belief back in the 1970s was that estrogen levels needed to be increased to help with PMS. While this did help relieve cramps for some women, it often did not help with the other symptoms and, for many women, it was actually harmful because their PMS symptoms were caused by already having too much estrogen in relation to progesterone. The pharmaceutical companies were not doing research on natural substances because they couldn't patent them. So the herb companies were operating with good intentions but little accurate information.

I had so many bad experiences with herbs that were supposed to be helpful for PMS or women's hormone problems that I became afraid to try the new ones that became available. But then—and this is before I knew about energy testing—I began to notice how my body reacted energetically to the different herbs on the shelf. My energy would literally separate from an herb that was not good for me; it would pull back. It was as if my energies and the energies of the substance didn't want to merge. That's how I energy tested before I knew energy testing. I would sense the energetic impact of the herb on me so that I wouldn't keep buying and ingesting substances that were harming me. Then came that fortuitous day when a young Mexican woman I knew was chewing on the root of a plant. I could see that there was an elegant energy between her and the plant. I asked her what it was. She explained to me that she did not feel well ("the time of the month," she confided) and that chewing on this root helped her immensely. The energy of that root felt so good to me that I asked her how I could get some. She generously cut off some of the root and gave it to me.

The next time I felt PMS symptoms, I began to chew the plant, Mexican yam root, which is very high in natural progesterone (Mexican and Siberian yam root are the best of the various forms of yam root I know about). It was so good for me that it was almost like a magic potion. But it tasted *so* bad. I talked a local health food store into grinding it into a powder so I could put it into little capsules that could be swallowed without having to taste it. I began checking each woman I knew who was having difficulty with PMS. I would put the yam root into her energy field to see if it made her energies look better. If it did, her PMS symptoms were related to low progesterone. Pretty soon, my friends and clients were getting the health food store to grind the root for them, and they were reporting such good results that the people at the health food store began asking their suppliers if they could get Mexican yam root in capsules. No one was distributing that at the time, but the Solaray rep took the idea back to his company and the people there decided to add it to their product line. Solaray sent me a year's supply with a letter thanking me for my contribution.

A few years later, in the late 1970s, Kathryn Dalton, a physician in England, did research with female prisoners who had been convicted of impulsive acts of violence. She suspected that these women were taken over by extreme PMS caused by a progesterone deficiency. After she provided them with progesterone supplements while they were premenstrual, their emotional imbalances and outbursts ceased. This was so obvious to everyone involved that she actually was able to get several women released from jail on the promise that they would faithfully take progesterone at the appropriate times.

When my father was dying, in 1987, a time of intense emotion for me, I found that yam root was no match for the extremes of my PMS. Taking a cue from Kathryn Dalton, I made it my quest to find pure, natural progesterone. As luck would have it, one of the very few pharmacies in the country that was able to supply natural progesterone was in my hometown of Ashland, Oregon. David wanted to nominate the pharmacist, Jack Sabin, for the Nobel Peace Prize.

However, as with all good things, the pendulum can swing too far. While I was viewing progesterone as God's gift to women, I assumed that if a little was good, more was better. I began to rely on progesterone any time I felt a little off emotionally. The result was that my uterus began to grow and thicken. That is progesterone's job, fortifying the uterus so it is able to carry a baby. While this started imperceptibly, it had become quite a problem before I made the connection between overdosing on progesterone and the changes in my uterus. I have been told that my uterus is the size it would be if I were two and a half months into a pregnancy. So here I am, stuck to this day with a tummy that won't go flat. The lesson is, everything in balance. And that is the great value of energy testing, to be able to assess what your body needs at any point in time. Meanwhile, I have been alarmed to see the culture's pendulum also swinging toward progesterone as a wonder drug. Credible health providers are advocating that everyone, with no assessment of their own natural progesterone/estrogen balance, should take progesterone in doses that are simply, according to my experiences, over the top.

Taking too much progesterone also creates a different ratio between your progesterone levels and all your other hormones, and this can have further unintended effects. I am someone who rarely gets depressed—I get hysterical—but suddenly everything was upside down. My levels of estrogen became too low in relationship to my progesterone, and I felt a deadness I'd never known before. When I backed off from taking too much progesterone, and particularly from taking it when I wasn't premenstrual, the deadness went away. In short, the progesterone panacea had its limitations. If I used it to keep all my emotions in control, it hurt my body, and eventually my spirit as well.

Over time, I learned that by combining energy techniques with the use of progesterone, each became more effective. PMS involves a *chemical* imbalance in hormones that can be corrected chemically *or* energetically, and it involves an *energetic* imbalance that also can be corrected using either approach. But working directly with your body's energies does not lead to unwanted side effects! And it can produce the same kinds of chemical changes for good feelings that ingesting hormones can produce.

For myself, when I had PMS, I could suddenly become hysterical. Most women know the signs that they are chemically, hormonally, or energetically off, and they can take steps to correct the imbalances using energy interventions. Family members can be supportive as well. David quickly learned his job. If he saw me become hysterical or anxious or extremely emotional, instead of derisively asking, "Oh, God, is it that time of the month again?!!" he would lovingly say: "Here, let me hold your triple warmer points." This was a form of first aid (not to mention self-defense) that would immediately calm me. Other steps could follow that would further balance and stabilize.

While the biochemistry of PMS is extraordinarily complex, if you approach it energetically, instead of somehow trying to control it chemical reaction by chemical reaction, it becomes manageable. And while the energies in your body can seem as complex as its biochemistry, what you can do to influence your energies is much simpler. Here is how it looks to me.

YOUR MONTHLY CYCLE: A PLAY OF MANY ENERGIES

When a new female client walks through the door, my eyes see the same things your eyes would see. In addition, I see a luminance surrounding her body, known in many healing traditions as the *aura* (page 30). If the woman is healthy and hormonally balanced, the energies surrounding her body seem full and billowy, and they go out a long way. If the woman is having a difficult premenstrual time, the aura will be collapsed, dense, and hovering close to her body. The energies that normally move in an easy exchange with the environment have a compressed look. Colors in the aura that normally move from one band to the next are splotchy. They do not flow smoothly. Beyond the surrounding aura, the energies that move *within* the body also change in their appearance at different times in her cycle.

When I look at the energies *in* a woman as she lies on my table, I might see any of many variations of how her energies are flowing. Emotions show up in the energies, and different emotions show up differently. With depression, the energies may appear to be sinking; with panic, it may seem they are rising, sometimes looking like they want to explode. If the woman is exhausted or ill, I might see merid-ian energy traveling backward, chakras (vortexes of energy) hardly spinning, or per-haps all the energy will be very slow. I may see blockages that need to be freed. Before the treatment is through, I hope to see the meridians streaming in a forward manner, the chakras spinning at a vibrant tempo, and the obstructions unblocked.

Hormonal imbalances also reveal themselves in the energies. The energies of a woman who is having difficulties with PMS often appear sluggish. There may be places where there is almost no movement, where energies are not reaching their natural destination. Sometimes still energies will move suddenly or change direc-tion or appear chaotic. They may seem compressed and wanting to burst.

Seeing such patterns in a woman's energies guides me. If we can bring her energies into balance and harmony, the woman is going to have a much easier pre-menstrual time. The energies will not, of course, look like her energies at non-premenstrual times, but they will not be as collapsed, compressed, or splotchy, and the flow among the colors will be smoother. Bringing about these changes in the energies will adjust the hormones. *Chemistry follows energy!* Simple energy proce-dures harmonize hormones!

The techniques I use are physical and they are simple. You do not need to be able to see the energies for the techniques to be helpful. They work whether you see the energies or not, and even whether or not you believe the techniques will work. That you do not need to be able to see energy for energy techniques to be ef-fective is, in fact, a core premise on which my approach to energy medicine is built. And I've seen this premise demonstrated again and again over the past 30 years.

ENERGY TECHNIQUES FOR MANAGING PMS

Any of the individual exercises in the Five-Minute Daily Energy Routine can serve as a kind of first aid to give you a boost during PMS. If you suddenly find your-self coming unglued, you can scream, you can hide, or you can do a Hook-up (page 62). If you feel vulnerable but are required to be with other people, do the Zip-up (page 61). The Three Thumps (page 53) instantly recharges your batteries. I am certain that the Wayne Cook Posture (page 56) helped save my marriage a num-ber of times by bringing me back to myself and my senses *before* I called the lawyer. It is one of my all-time favorite unscramblers. Use it whenever you are feeling stressed or overwhelmed. Another technique that is invaluable when you can't think clearly, your physical coordination is off, or you find yourself spiraling into depression (any of which is a sign that you are homolateral) is the Homolateral Crossover (page 67). In fact, I know of nothing else that works as well when your energies are not crossing over from one brain hemisphere to the other. If you can't get yourself into gear, if nothing else seems to be working, assume that you need it.

The energy techniques for managing PMS are grouped as follows: (1) Maintaining good energies throughout the month; (2) Calming steps to take at the first signs of PMS; (3) Adding a "PMS Module" to your Daily Energy Routine; (4) Techniques for working with physical discomfort and pain; (5) First aid for the emotional crazies; and (6) Energy testing for hormonal and other supplements.

1. MAINTAINING GOOD ENERGIES THROUGHOUT THE MONTH

Energy habits are the established ways your energies flow, and they may be for better or for worse. If you can institute positive energy habits in your body during times that are less stressful, your body will be able to cope more effectively when stress hits. Keep these principles from Chapter 2 in mind:

- Stretching makes space in the body so your energies can move in their most natural way.
- Clearing toxins supports the healthy flow of your body's energies.
- Energies need to move in crossover patterns.
- A set of simple exercises done on a daily basis can stimulate all your energy systems, which in turn brings a better vibration to your organs, your immune system, and your spirit.

By doing the Five-Minute Daily Energy Routine daily and combining it as needed with other stretching and crossover exercises from Chapter 2, you will be helping your body establish healthy energy habits that will better support you throughout your monthly cycle. Beyond that, time will show that these healthy energy habits become embedded—so you are giving yourself a gift that keeps on giving.

Most women intuitively know that they are best served by going internal during the premenstrual part of their cycle. Cooperate! Do not schedule nonessential social activities. Plan for moon-hut time. Keep life as calm and simple as possible. And be prepared. If the miseries of PMS occur, be ready to swiftly shift your energies. The following techniques show you how. I suggest that you familiarize your-self with them during the time of the month when you are not premenstrual so that you can use them with a minimum of effort. If you learn them during calmer times, you are much more likely to use them when you need them.

The techniques in the remainder of this chapter, and in the remainder of this book for that matter, can seem like so many more chores added to your day. The thing is, they work. And they do not take long. They take less time than a brisk walk or most people's driving time to the gym. In these ways, they are very effi-cient. Also, these exercises are much easier to do than they seem at first when read from the printed page. And you need to do only a portion of them to get helped. I try to give you good guidance in determining which ones are going to be the most useful for you.

If you treat this book as a guide, you will have at your disposal powerful, simple tools for being more in charge of your life, more at peace, and more able to deal with life's challenges. I hope you will come along with me on this journey and give the techniques a fair try. So begin with the routines in Chapter 2, and then when you notice the first symptoms of PMS, take the following definitive steps.

2. CALMING STEPS TO TAKE AT THE FIRST SIGNS OF PMS

At the first bout of irritability, lethargy, bloating, or other sign of PMS, carve out some time and make yourself very comfortable. My preference is to be in a bath-tub or hot tub. Immersing our bodies in warm water may be the closest experience we have to returning to the womb. It shifts your energy field so that your body is more receptive to healing and energy exercises that calm your hormones. Once you have relaxed a bit (lovely soothing music in the background enhances the experi-ence), lie back in the tub, get completely comfortable, and do the following:

A. Sedating the Triple Warmer with the Smoothie (time—20 seconds)

You were introduced to this procedure in the previous chapter. When you are stressed, whether due to external circumstances or bodily disruptions such as those that are part of PMS, triple warmer steps into action. Triple warmer, you may recall, is designed to keep you safe by invoking the immune or fight-or-flight re-sponse and by maintaining survival habits. But because it evolved for a world that no longer exists, it is often on overdrive. During PMS, it is almost always valuable to calm the triple warmer energies that move through your body. This relaxes your entire energy system, releases tension, and reduces fear and anxiety. Fortunately, the Triple Warmer Smoothie (page 107) is a quick and simple way to calm triple warmer, and sometimes it is all that is needed. The Triple Warmer Tap (page 108) and the Triple Warmer/Spleen Hug (page 110) are also simple and effective. A more decisive though more time-consuming approach is to hold the Triple Warmer Sedating Points (page 110).

B. Holding the Spleen Meridian Strengthening Points
(time—about 5 minutes)

Spleen and triple warmer meridians exchange energy in a teeter-totter kind of relationship. When triple warmer is activated in PMS (or in any other situation), it springs into action and drains energy from spleen to nourish itself. Spleen meridian often gets so depleted that your immune system and your ability to adapt become compromised. When spleen meridian energy is low, your body insists that you slow down. If you go with this slower flow, you can do much to restore balance. By *deliberately* slowing down in the bathtub or another comfortable setting, and then calming triple warmer and strengthening spleen meridian, you are telling your body that you are attuning yourself to the monthly shift that is about to occur. You are speaking to your body in its native language—*energy.* This prepares you to usher in this period attuned to the needs of your body, mind, and spirit. Strengthening spleen meridian also happens to help with the heavy, weak, and aching legs that are often a symptom of PMS.

Meridians can be strengthened in several ways. One of the most powerful is by holding acupressure points that affect the meridian. The procedure is similar to the way you sedated triple warmer in the previous chapter. You simply place your fingers on the points, hold for a couple of minutes, and then hold a second set of points. By finding a comfortable position in the tub, on the couch, or in another setting, this can be an enjoyable, centering, meditative time. While that is my favorite way to do it, you can also multitask. Holding the points also works while you are watching television, talking on the speakerphone, or chatting with a friend. You can even get your friend to hold the points for you (adjust the instructions accordingly).

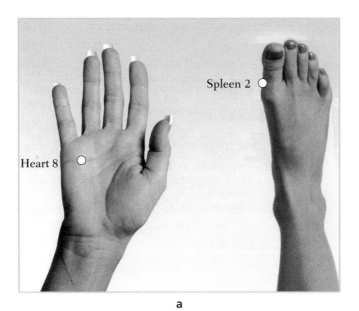
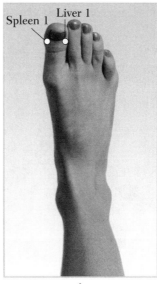

Figure 4-1
SPLEEN MERIDIAN STRENGTHENING POINTS

To strengthen spleen meridian, simultaneously place your fingers on Speen 2 and Heart 8 on either side of your body and hold for about two minutes (see Figure 4-1a). You can do this by placing the middle finger of one hand on Heart 8 and the middle finger of the other hand on Spleen 2. Repeat on the other side. For the second set of points (Figure 4-1b), use the thumb and pointer fingers to hold Liver 1 and Spleen 1. You can hold these points on both feet at the same time. Hold for about a minute.

3. ADDING A "PMS MODULE" TO YOUR DAILY ENERGY ROUTINE

To keep your energies humming during your premenstrual time, you might want to do the Five-Minute Daily Energy Routine twice, once early in your day and again in the afternoon or evening, particularly when you are dragging. I suggest adding the following techniques to the routine when you are premenstrual and continuing to include them until two or three days after your period begins. Insert the following five techniques right after the Lymphatic Massage (page 59), and then end the routine as usual, with the Zip-up and Hook-up.

A. Connecting Heaven and Earth (time—about 2 minutes)

This is an ancient exercise that women have used in many cultures throughout time. See instructions on page 44.

Figure 4-2
THE TRIPLE AXIS HOLD

B. The Triple Axis Hold
(time—about 30 seconds)

The Triple Axis Hold is a quick and simple technique that brings immediate comfort. It calms and balances your hormones, and benefits the pituitary and hypothalamus glands in particular:

1. Place the heel of either hand on your forehead, then take your middle finger and rest it on the top of your head.
2. Make a three-finger cluster with the thumb, index, and third fingers of your other hand and place it just beyond the curve of your head (see Figure 4-2).
3. Hold during three deep breaths.

C. Abdominal Stretch (time—about 30 seconds)

Your abdomen, with all the organs in it, can become full and tight before your period, interfering with the normal movement of energy. This bloating causes a feeling in the middle of your torso like a dam being backed up. The Abdominal Stretch opens the spaces, relieves some of the pressure, and allows fresh energy to move through the area, giving you a reprieve from that unpleasant stuffed-bird feeling.

1. Stand facing the back of a chair or other piece of furniture that is a similar height, placing your hands on top of it. Straighten your arms. Alternatively,

reach to a ledge about the height of your shoulders and straighten your arms, standing now an arms distance from the ledge.

2. Look up, and while stretching your neck gently backward, swing your right leg backward, keeping it straight at the knee. Feel the stretches in your neck and your abdomen (see Figure 4-3).

3. Return to center.

4. Repeat with your left leg.

5. This will feel good. Do as many as you wish.

Figure 4-3
THE ABDOMINAL STRETCH

D. The Sideways Stretch
(time—less than a minute)

This exercise also stretches the ligaments in your abdomen and is particularly beneficial for the energetic functioning of your heart, liver, spleen, and gallbladder, each of which may become stressed during your premenstrual time. Like the Abdominal Stretch, it also helps counteract bloating. To do the Sideways Stretch:

1. Stand with your hands on your thighs and take a deep breath, inhaling through your nose and exhaling through your mouth.

2. Swing your arms out to the side and above your head on a deep in-breath.

3. Grab your right wrist with your left hand, and on an out-breath, bend to the left while pulling your right arm with your left hand (see Figure 4-4).

4. Return to center on an in-breath. Repeat on the other side. Do at least three rounds.

Figure 4-4
THE SIDEWAYS STRETCH

135

5. Finally, with palms facing outward and hands over your head, circle your arms down to your sides.

E. Spleen Meridian Flush and Tap (time—less than a minute)

If you feel weak during your premenstrual time, spleen meridian almost always needs to be strengthened. A very good way to strengthen a meridian is by flushing it, which is to trace it backward one time and forward three times, kind of like backwashing a filter, pulling out stagnant energies. You can brush the meridian lightly with your hands or keep your hands a couple of inches above the meridian, using a slow, deliberate motion. You will be following the meridian pathway shown in Figure 4-5. Throughout the exercise, move your hands slowly and deliberately. To flush and tap spleen meridian:

1. Stand with your hands flat on the sides of your body at your waist, fingers pointing straight down.
2. Take a deep in-breath, and draw your fingers up to your armpits.
3. Let your breath out as you bring your hands down your sides, again flattening them. When you reach your waist, move your hands to the front of your hipbones, then continue down the inside of your legs with fingers spread, hands flat.
4. Go over your anklebones and continue forward along the insides of the feet and off the big toes.
5. Take a deep in-breath and retrace these steps in the other direction, starting at the insides of the feet at the big toes and, with your hands flat, trace straight up the insides of the legs, flaring out at the hips, to the sides of the body, up to the armpits, and then down to the bottom of the rib cage. Retrace in this direction two more times.
6. Tap the spleen neurolymphatic reflex points beneath the breasts, in line with your nipples, and down one rib for about ten seconds as you breathe deeply (see Figure 2-5, page 54).
7. Slide your fingers to the spleen acupressure points on the side of your body and tap for ten more seconds (see Figure 4-5).

Figure 4-5
SPLEEN FLUSH AND TAP

∞ 4. TECHNIQUES FOR WORKING WITH PHYSICAL DISCOMFORT AND PAIN

Among the most common physical symptoms of PMS are bloating, breast tenderness, cramps, back pain, and groin pain. For cramps and many types of pain, people often instinctively circle their hand over the area of discomfort. There is good reason to do this. Your hand can pull blocked or stagnant energy out of an area. Circling in a counterclockwise motion tends to draw out disruptive energy; circling in a clockwise motion tends to stabilize the energies that are there. Your hand does not even need to make physical contact with your body. Circle a couple of inches above the area of discomfort in a counterclockwise motion and notice if you feel your hand pulling energy off your body. Shake your hand off from time to time, as if sending the excess energy into the ground. Once you are feeling even a little bit better, you can stabilize the improvement by circling in a clockwise direction.

A. Stomach Meridian Flush (time—about a minute)

Another good technique for PMS discomfort is to flush the stomach meridian. Rather than taking energy out of the specific area of discomfort (as with circling), tracing the stomach meridian backward (the first part of flushing the meridian) is like cleansing a pathway that keeps the energies involved with PMS flowing. It immediately relaxes everything in the area of your stomach, so it is a natural response to cramps or stomachaches. But it can also help with difficult emotions, particularly heartache or other emotional pain whose physical counterpart is in your stomach or chest.

As with flushing the spleen meridian, you can brush the meridian lightly with your hands or keep your hands a couple of inches above the meridian, using a slow, deliberate motion. Begin by tracing the stomach meridian backward, using both hands to follow the path of the meridian in the opposite direction of its normal flow (see Figure 4-6):

1. Starting with your fingers at the second toe of each foot, travel to your ankles and straight up your body, slightly toward the outsides of your legs.
2. Come straight up the front of your body, over your ovaries, over your breasts, straight up your neck, over your jawbone, and up through your eyes, to your hairline.

Beginning of the
Stomach Meridian

End

Figure 4-6
STOMACH MERIDIAN FLUSH

3. Then circle down, following the outline of your face, down to your jaw-bone, and finally come straight up to the points beneath your eyes.

4. Shake the energies off your hands. Trace this backward pathway again one or two times if you wish.

5. Now trace the stomach meridian in the opposite (forward) direction (again, see Figure 4-6).

6. Trace the meridian forward a total of three times. End by shaking the energies off your hands.

B. Additional procedures to address specific types of PMS discomfort

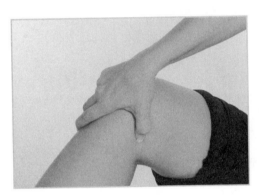

Figure 4-7
Massaging Spleen 9

Bloating. The best exercises I know for bloating are the Abdominal Stretch and the Sideways Stretch, both suggested as additions to your daily routine. Any stretch that involves the stomach opens space for energy to move. You can also massage an acupressure point called Spleen 9 on the inside of the leg beneath the knee, in the depression behind the tibia (see Figure 4-7) for about ten seconds in a counterclockwise direction (as if your knee were the clock), and then tap it for five seconds. This is a traditional treatment for edema, though it is not suggested for women who are pregnant.[1]

Breast Swelling and Tenderness. PMS pain and swelling are signs of hormonal changes that block the movement of energy, causing excess energy to accumulate and become dense and sluggish. Your stomach meridian passes over your breasts, down through your abdomen, and over your ovaries. During PMS, the stomach meridian is often overtaxed, trying to manage the complex changes occurring in these areas. It often accumulates excess energy. When your breasts swell or are tender, releasing some of that energy can provide relief. Begin by tracing the stomach meridian backward as described in the discussion of the Stomach Meridian Flush. The sedating stomach meridian can further reduce the pressure in these areas, giving relief to cramps as well as breast tenderness, and pain in the ovaries. Sedating the meridian is like opening a tap that allows this energy to drain and allows healthy circulation to be restored. Instructions for sedating the stomach meridian, and several other meridians involved with PMS, are presented after the following discussion of cramps, back pain, groin pain, late periods, and sedating the meridians.

Cramps. Sedating the stomach meridian also helps with cramps. A procedure that specifically focuses on cramps is to massage the fourth acupressure point on the spleen meridian. Relax, position your feet so you can comfortably reach both, and simultaneously massage the points that are on the inside of each foot, halfway between your toe and your heel (Spleen 4, see Figure 4-8). Push up against the bone and massage in a counterclockwise direction for at least ten seconds and up to a minute. Then tap the points while taking two deep breaths. Meanwhile, massaging Liver 4 (also shown in Figure 4-8) relieves pressure from the uterus and

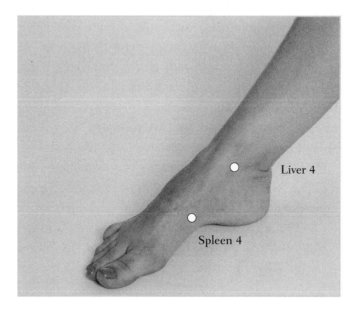

Figure 4-8
CRAMP RELIEVER

may also provide relief from cramps. Again, massage in a counterclockwise direction and then tap the points during a couple of deep breaths.

Back Pain. In energy medicine, pain relief is based on addressing the underlying problem rather than masking the pain, as with "painkillers." Instead of just blocking out the symptoms, energy medicine techniques for alleviating pain correct, at the level of the body's energies, the reasons for the pain. PMS can cause pain in a sturdy back because the kidneys have to filter more blood and hormones during this time and can become full and stressed, leading to pain in the surrounding area. Sedating the kidney meridian removes tense energy from the kidneys, relaxes them, and provides relief. If you have upper back pain together with cramps, sedating the small intestine meridian, which governs the entire abdominal cavity, can be helpful. This can relieve cramps as well as back tension. When back pain is located in and around the spine at the waist, the large intestine meridian is often involved, and sedating the large intestine meridian can provide relief.

Groin Pain. During PMS, as the uterine lining thickens, the weight of the uterus increases. This places tension on the groin ligaments, which may tighten in order to give the uterus better support but can result in groin pain. The simplest technique is to lie down and place your fingertips right on the groin lines where your legs attach to your body. Relax your hands so they are lying flat. Your hands are bringing good energy to your ovaries, the sides of your uterus, and two valves (the ileocecal and Houston's valves) that often lose their rhythm during PMS, causing constipation. Hold this position for about two minutes.

Another approach for relieving groin pain is to sedate with the liver meridian. This meridian controls the tension in the ligaments of the body, and the liver itself has the job of dispelling the hormones that have been released into the system and need to be eliminated. So the liver meridian is also working overtime during PMS. Sedating the liver meridian (described on page 146) can relax unnecessary tension in the ligaments, help the system restabilize itself, and help the liver process hormones that have served their function.

If Your Period Is Late. You can often get your period to start by relaxing the gluteal muscles and the entire pelvic area. Sedating a meridian that controls the circulation system, the pericardium, and the gluteal muscles—called the circulation-sex meridian—can bring on a late period with no risk of aborting a fetus. Another reason keeping this meridian in a good flow is so important is that it works synergistically with the spleen, stomach, and triple warmer meridians.

 ### Sedating the Meridians Involved with PMS

Many of the uncomfortable symptoms associated with PMS are based on excess energy rather than not enough energy. This may seem counterintuitive because you may be feeling that you need more energy, which is also true. Here is how it works. When too much energy accumulates in certain areas of the body, the normal flow of energy is blocked, causing weariness and pain. When there is too much energy in one place, the flow stops, so there is too little energy getting to other areas. The reason too much energy may accumulate in an area is that when a meridian is being overworked (as happens with PMS), the meridian may overcompensate, or other meridians may feed it extra energy in an effort to support it.

A powerful way to help get blocked energy flowing again is to "sedate" the meridian pathway in which the clogged energy has accumulated. This makes it possible for the energy to start moving again and can provide tremendous relief while allowing your whole system to get back into balance. The procedure is similar for each of the meridians, though different acupressure points are involved with each. In Chapter 3, you were shown how to sedate the triple warmer. Again, when you sedate a meridian, you place your fingers on the designated acupressure points, hold for a couple of minutes (it is a firm touch but does not involve deep pressure or massage), and then hold a second set of points for about a minute. The total time involved is three to six minutes, depending on whether it is physically possible to hold both sides simultaneously. Be sure when holding the points to find a comfortable position so you are not taxing your body. The meridians most involved with the *physical* symptoms of PMS include the stomach, small intestine, kidney, large intestine, liver, and circulation-sex meridians. The earlier discussion of specific PMS symptoms tells you when it might be valuable to sedate each.

Sedating the Stomach Meridian (breast tenderness and cramps; time— about five minutes)

You can work from the figures, holding the first set of points (Figure 4-9a) for two minutes on one side, then the other side; and the second set of points (Figure 4-9b) simultaneously, for about a minute. Here are the instructions spelled out in greater detail:

1. Take the thumb and middle finger of your right hand and hold the end of your right second toe (this covers the Stomach 45 acupressure point).
2. At the same time, hold the sides of the end of the index finger of your right hand with the thumb and index finger of your left hand—this covers Large Intestine 1 (see Figure 4-9a).
3. Hold for about two minutes. Be sure you are physically comfortable.
4. Repeat on the other side.
5. Bring your feet into a position where you can comfortably reach both with your hands.
6. Move your thumbs along the valley between the ligaments of the second toe and third toe of each foot until you are about an inch back from where your toes begin (Stomach 43).
7. Take the index fingers of each hand and place them on the ligament that

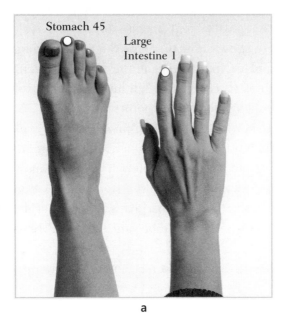

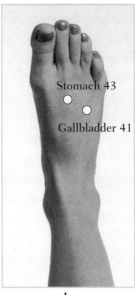

a   b

Figure 4-9
STOMACH MERIDIAN SEDATING POINTS

goes to your fourth toes, about an inch farther back (toward your ankle) from the position of your thumb (Gallbadder 41) [see Figure 4-9b].

8. Hold for about 60 seconds.

Sedating the Small Intestine Meridian (pain in upper back combined with cramps; time—6 minutes)

Simultaneously place the middle finger of one hand on Stomach 36 (a hand's width beneath your knee, in line with your second toe) and the middle finger of the other hand on Small Intestine 8 (about three fingers' width beneath your elbow, in line with the little finger), on either side of your body, and hold for about two minutes (see Figure 4-10a). Repeat on the other side. Then place the middle finger of one hand on Small Intestine 2 (beneath the little finger, between the base of the finger and the wrist) and the middle finger of the other hand on Bladder 66 (in the indentation at the outside bottom of the little toe) and hold for about a minute (see Figure 4-10b). Repeat on the other side.

Sedating the Kidney Meridian (back pain; time—about 3 minutes)

It is possible to hold both sets of points simultaneously with this one.

1. To do both sides at once, sit in a chair and pull your feet up onto the chair. Place the middle finger of each hand between the second and third toes and bring these fingers over the ball of the foot, resting them on the edge of the ball that is away from your toes (Kidney 1). You can hold either foot with your right or left hand, whichever is easier.
2. Place your thumbs on the base of the opposite big toenail, in the corner closest to the second toe (Liver 1) [see Figure 4-11a].
3. Be sure you are comfortable. Hold for about two minutes. If you are pregnant, however, hold for only about 30 seconds.[1]
4. Place the middle finger of your right hand on the outside of your right big toe and slide just over the "bunion bone" to the indentation (Spleen 3). Find the same point on your left foot.

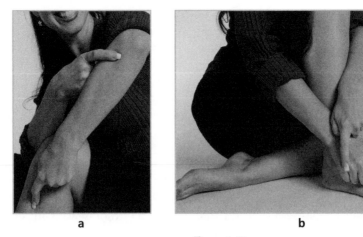

a   b

Figure 4-10
SMALL INTESTINE MERIDIAN SEDATING POINTS

5. Place your thumbs in the indent just behind your inner anklebones (Kidney 3) [see Figure 4-11b].

6. Hold about 60 seconds.

Sedating the Large Intestine Meridian (pain in lower back; time—about 6 minutes)

Figure 4-12a shows the first pair of points, Large Intestine 2 (inside the base of the index finger) and Bladder 66 (outside the base of the little toe, in the indentation). Hold for about two minutes on each side. Figure 4-12b shows the second pair, Large Intestine 5 (on top the of wrist in line with the index finger) and Small Intestine 5 (on top of the wrist, in line with the little finger). Hold for about one minute on each side.

Sedating the Liver Meridian (groin pain; time—about 6 minutes)

Simultaneously place the middle finger of one hand on Liver 2 and the middle finger of the other hand on Heart 8 (see Figure 4-13a), on either side of your body, and hold for about two minutes. Repeat on the other side. Then place the middle

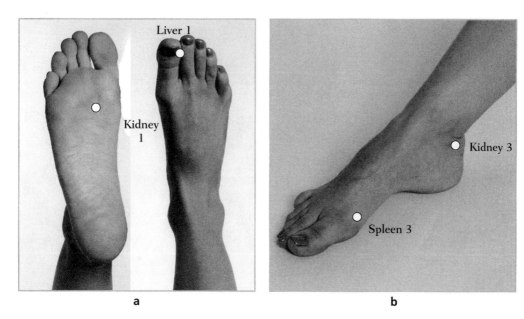

a   b

Figure 4-11
KIDNEY MERIDIAN SEDATING POINTS

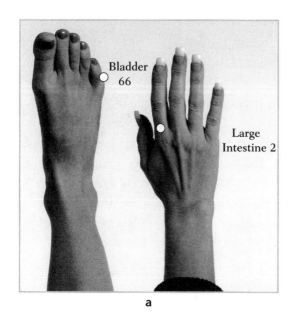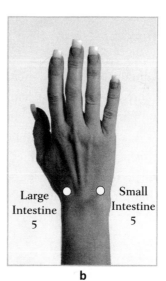

a   b

Figure 4-12
LARGE INTESTINE MERIDIAN SEDATING POINTS

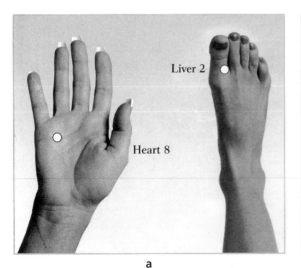

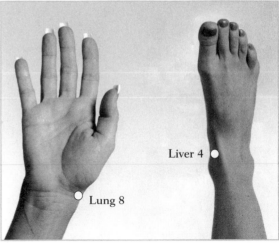

a   b

Figure 4-13
LIVER MERIDIAN SEDATING POINTS

finger of one hand on Liver 4 and the middle finger of the other hand on Lung 8 (see Figure 4-13b) and hold for about a minute. Repeat on the other side.

Sedating the Circulation-Sex Meridian (late period; time—about 6 minutes)

Simultaneously place the middle finger of one hand on Spleen 3 (inside of the foot, just before the lump that leads to the big toe) and the middle finger of the other hand on Circulation-Sex 7 (middle of the wrist, in line with the middle finger), on either side of your body, and hold for about two minutes (see Figure 4-14a). Repeat on the other side. Then place the middle finger or thumb of one hand on Kidney 10 (on the inside of the leg where the knee bends) and the middle finger of the other hand on Circulation-Sex 3 (on the inside of the elbow, in the middle) and hold for about a minute (see Figure 4-14b). Repeat on the other side.

∞ 5. FIRST AID FOR THE EMOTIONAL CRAZIES

Energy medicine offers a number of techniques worth having at your fingertips during any emotional roller coaster, and PMS can be the Giant Dipper of emo-

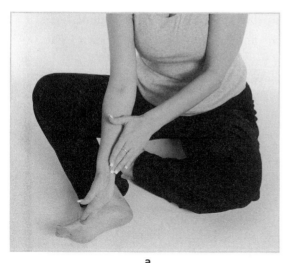

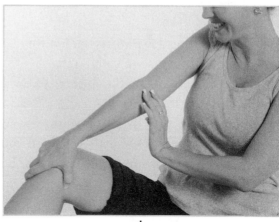

a   b

Figure 4-14
CIRCULATION-SEX MERIDIAN SEDATING POINTS

tional roller coasters. I can assure you of several things about energy medicine methods for emotional turmoil. They are quick. They are powerful. You can learn them in advance so you have instant access. You will feel a shift in the energy. And you will not remember to use them when you need them. Okay, prove me wrong!

Three principles to keep in mind are stretch, tap, and cross and hold. If you don't remember anything else, breathe deeply as you: (1) stretch your midsection, (2) tap anywhere on your face, head, and body, and (3) cross your arms and hold your shoulders. You have also already learned several great emotional first-aid techniques: the Hook-up (page 62), Zip-up (page 61), Triple Warmer Smoothie (page 107), and Wayne Cook Posture (page 56). Four others you can experiment with are Charging Your Batteries, Expelling the Venom, the Neurovascular Hold, and the Darth Vader Breath.

A. Charging Your Batteries (time—a little more than a minute)

When you feel exhausted, on edge, or ready to burst into tears, this technique can bring relaxation, support, and strength to the liver, spleen, bladder, and kidney meridians, each of which can be highly stressed when you have PMS. If done regularly, it will help keep your hormones balanced and bring greater ease to your cycle.

1. Sitting down with your left foot bent in your lap, squeeze the sides of your left foot (see Figure 4-15a).
2. Begin to drag the fingers of your left hand with pressure up the inside of your leg.
3. Draw your hands up your torso, with pressure, your left hand moving up the left side of your body and your right hand moving up the left side of the midline at the front of your body.
4. When your right hand reaches underneath your collar bone (K-27) and your left hand reaches to the side of your breasts on your rib cage (Spleen 21), tap both areas 10 or 12 times while breathing deeply (see Figure 4-15b).
5. Repeat on the other side.

a   b

Figure 4-15
CHARGING YOUR BATTERIES

B. Expelling the Venom (time—less than a minute)

The first two parts of the Blow-out/Zip-up/Hook-in technique from Chapter 2 (page 74) are so powerful that in tandem, I call them Expelling the Venom. They move stress and rage out of your body, unblock congested energies, and free the spirit:

1. Stand up straight. Put your arms out in front of you, bend your elbows slightly, make fists with the insides of your wrists facing up, and take a very full breath (see Figure 2-16, page 75).
2. Swing your arms behind you and up over your head. Hold here for a moment.
3. Reach way up, turn your fists so your fisted fingers are facing each other, and rush your arms down the front of your body as you emphatically release your fists. Let out your breath and your emotions with a "whooooosh" sound or any other sounds that come naturally.
4. Repeat several times. This will feel good. The last time, bring your arms down in a slow and controlled manner, blowing your breath out of your mouth as you go.
5. End with a Zip-up. With a deep breath, place your hands at your pubic bone (see Figure 2-10, page 61) and pull them straight up the midline of your body, up in front of your face. Stretch high above your head, and turn your palms outward. Slowly releasing your breath, extend your arms out to the sides as far as they will go and back down to your pubic bone.

C. The Neurovascular Hold (time—1 to 3 minutes)

When you are stressed or feeling blue, this technique releases tension from your body as well as from your mind. Your body still reacts to threat and stress as our ancestors' bodies did millions of years ago. Running away or fighting required that the blood rush away from the brain and into the body to support quick action. You are designed to *act* in an emergency, not think. The Neurovascular Hold, already mentioned in the previous chapter, interrupts the fight-or-flight response, bringing blood back to your forebrain so you can think more clearly.

1. Sitting or lying down, tune in to a stress you are feeling or focus on a stressful thought, memory, or situation.
2. Place your thumbs at your temples and the pads of your fingers on your forehead, on the bony area above your eyebrows (see Figure 3-1, page 106).
3. Hold here softly for up to three minutes, breathing deeply, in through your nose and out through your mouth.

Alternative: Place the palm of one hand across your forehead and the palm of your other hand across the back of your head. Again, hold softly for up to three minutes, breathing deeply.

D. The Darth Vader Breath (time—less than a minute)

This technique calms triple warmer, balances hormones, releases stress, and brings a steadiness to body and mind. The breathing for this technique is very slow and controlled.

1. Stand and empty your breath completely.
2. Breathe in through your mouth, pulling your breath very slowly through the back of your throat, making a raspy "Darth Vader" type of sound.
3. Breathe out making the same sound, again very slowly.

While some of these methods will work better for you than others, each can be effective. Experiment.

6. ENERGY TESTING FOR HORMONAL AND OTHER SUPPLEMENTS

While our emphasis has been on techniques for shifting your body's energies, a definition from Chapter 3 provides the context: Energy medicine is the art and science of working with the relationships between energy *and chemistry* to promote health, vitality, and well-being. Our focus is on energy because, in addition to that being my specialty, it is the area that has received less concrete attention in conventional as well as alternative approaches to helping women with their health concerns. But in practice, energy medicine is very attuned to chemical as well as energetic balances and imbalances. All herbs and medicines, in fact, have energetic properties. And energy testing (see pages 63–65 and the Appendix) is a tool for determining the types of hormonal, nutritional, or other supplements that might best complement hands-on energy interventions.

If you know that you always feel better after taking a particular herb to relieve your cramps, there is no need to energy test it. But energy testing can help you decide which of the eight herbs at the health food store is most likely to work for your unique body, complaints, and energy system. Every food, pill, or capsule carries an energy field. Energy testing allows you to determine, before ingesting it, whether the unique vibration of that substance resonates and is compatible with your own body's energies and needs. Since chemistry follows energy, that will also give you an indication of how your body will respond to the substance.

For myself and my approach with clients, I first try to handle monthly fluctuations using the energy techniques alone because they are less invasive. If you rely on outside sources for a chemical your body produces naturally, your body's ability to make that chemical is often diminished (an important exception is progesterone).

Energy techniques certainly work independently of any other interventions. But there are times where herbs and other supplements that are used for PMS can complement energy work, and I fully support that combination when it is necessary.

Even if you do not take it upon yourself to learn energy testing, you can still experiment with the substances described here on a trial-and-error basis. I have personally learned a great deal about my own body and what it needs through such experimentation, particularly before I was introduced to energy testing in the late 1970s. Please check the Internet for contraindications and discuss with a competent practitioner before ingesting any substance about which you are not well informed.

If you have cultivated an ability with energy testing, or have access to a practitioner who can reliably energy test substances you are considering, you will not only depend less on trial and error, you will also be able to figure out in advance the amount you need and when you need it. My body may need a tad of progesterone, but three tads may throw off my progesterone-estrogen balance. The next day I may not need any, and the day after that I may need three tads in the morning and one tad in the afternoon. I know of no better tool than energy testing for establishing a harmony between your body's fluctuating chemistry and what you offer it to maintain peak functioning.

Monthly changes in estrogen and progesterone balances are at the core of the most severe PMS symptoms. Estrogen levels are at their peak right up to ovulation, when the hormone is telling you to find a suitor and get on with it. With ovulation, progesterone levels begin to rise. If you did the first part right, according to nature's calling, then your body needs to start preparing for a baby. Progesterone remodels the uterus to carry the baby. Whereas estrogen excites you and makes you more sexually active and interested, progesterone calms you. It makes you more tranquil. In fact, one of its primary functions besides strengthening the uterus is to relax the central nervous system. But if you are in the postovulation part of your cycle and not producing enough progesterone, there is a deep-body anxiety that your uterus is not able to build the lining it needs for the anticipated baby—something is not right! Beyond this, your entire nervous system is out of balance because the expected progesterone is not being produced.

Improper progesterone levels may be due to many causes. Nature's design lottery simply edged some women toward more estrogen and others toward more progesterone. Many women begin to have a decline in their natural progesterone levels by their mid- to late thirties. If PMS is, for you, a time of increased anxiousness, hysteria, or reactivity, you may not be producing as much progesterone as your body needs. While some of the energy exercises in this chapter may lead to a better balance, you might also consider supplementing what your body is producing. Other signs of low progesterone for menstruating women, beyond irritability and extreme mood swings, include irregular menstrual cycles, heavy menstrual bleeding, bloating, back pain, edema, fatigue, excessive food cravings, endometriosis, uterine fibroids, and tender breasts.

While the most terrible parts of my own PMS were emotional, my younger daughter, Dondi, was physically debilitated every month, going into intense sweats and shakes, with terrible cramps and pain. And she never knew when her period would begin, so she could not plan for it, which proved costly for her in her career as an actress. Whereas progesterone was my wonder drug, it did nothing for her. The surprise solution was when she started taking birth control pills, which increased her estrogen levels. Suddenly and decisively, her period was regular and her symptoms disappeared.

For supplemental progesterone, the simplest way to introduce it into your body is to take Mexican or Siberian **yam root** capsules. Some people also use yam root or other progesterone creams. All of this is available over the counter. The strongest forms of progesterone are by prescription only (be sure you get natural progesterone rather than progestin, which is synthetic and more likely to be harmful). Natural progesterone may be taken as a suppository or orally in capsules, or you can punch a hole in the capsule and spread small amounts on your inner wrist or abdomen, allowing you to completely control the dosage. Your skin will absorb it directly into your bloodstream, whereas if you take it orally, your liver needs to process it. From yam root to prescription progesterone, you can experiment and see what happens, or you can energy test to find out which form and how much your body needs day by day. Sometimes that need fluctuates hour by hour. And in any case, please be careful not to make the error I made of seeing progesterone as a panacea and using it so much that it caused harm (see page 127). Possible symptoms of excess progesterone include irritability, dullness, depression, diminished libido, uterine cramping, headaches, and insomnia.

Before I found yam root, **vitamin B$_6$** was the only substance that really helped regulate my extreme PMS mood swings. B$_6$ calms the nervous system. I took it in higher doses than are generally recommended, but again you can energy test. I use **B complex,** which helps process the hormonal fallout of the stresses of daily life, along with extra B$_6$. **Potassium,** or sometimes a full-spectrum mineral capsule, helps normalize blood pressure and maintain electrolyte balances, which are also important for the nervous system. **Magnesium** is a mineral that is often deficient in women who have PMS. Magnesium relaxes the muscles, so it can help with numerous symptoms, from abdominal and leg cramps to headaches to constipation. Another favorite of mine was **chromium,** which helps balance blood sugar levels. Many women have sugar and chocolate cravings when they have PMS. Chocolate cravings often reflect the body's need for magnesium; sweet cravings, its need for chromium.

Many women experience a weakness in their legs during their premenstrual time. An herb called **butcher's broom** was a godsend to me when my legs would get weak. I could pop it before I danced or climbed in our Oregon mountains. It allowed me to use my legs in ways I couldn't have imagined while they were feeling weak and achy. The derivation of the name butcher's broom is interesting. During the plague in Europe, butchers were not afflicted. It turned out that butchers made special brooms of a shrubby evergreen plant, which absorbed into their hands (particularly where they had cuts), and this plant turned out to have strong medicinal properties. Butcher's broom works by attaching to the collagen and strengthening the blood vessels, which helps keep blood circulating throughout the body.

Other herbs that are frequently recommended for PMS include motherwort, dandelion root, dong quai, and ginkgo biloba. You can experiment with these or energy test, and you can learn a lot more about each of them on the Internet. Briefly, **motherwort** is said to calm the emotions and bring inner security. **Dandelion root** is used to strengthen the spleen and liver meridians as well as to help with water retention. **Dong quai** balances estrogen and progesterone levels for some women, though it always made me worse, another reason for energy testing. **Ginkgo biloba** is used to help with memory, brain fog, and blood circulation—I can't recall if it worked for me. Increasing the **fiber** in your food can help keep your digestive system moving during sluggish times. **Full-spectrum lights** give some women an energy boost and that uplifts their mood.

Your monthly cycle, however challenging, is a truly miraculous process. My hope is that the simple energy techniques described in this chapter will give you tools for meeting those challenges more effectively, and that they will help you tap more readily into the sources of deep wisdom that can be accessed with every cycle of the moon.

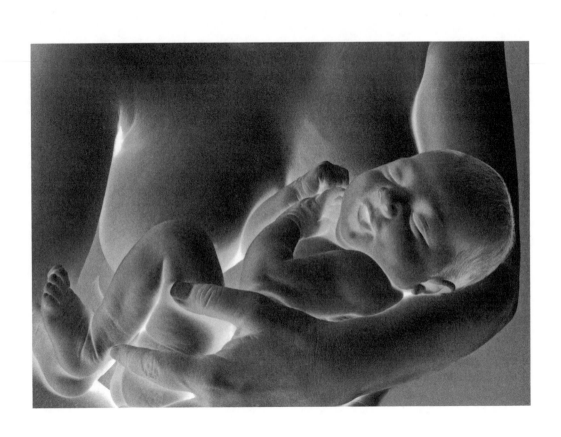

Sexuality, Fertility, Pregnancy, and Birth

*For centuries, women stood in their caves stirring the soup with
one hand, bouncing a baby on the hip, and kicking the woolly mammoth
out the door with the other foot. Women have, over the centuries,
developed a brain structure and style of thinking that is
characteristically different from men.*

—Jean Houston, Ph.D.

A human, according to a somewhat reductionistic accounting, is "a gene's way of making another gene." We experience the process more lavishly. Our sexuality is juicy, ecstatic, and compelling; our fertility, the pinnacle of potential; pregnancy, the journey from a microdot to a miracle; birth, the exemplar of manifestation.

Our bodies know the plan. Our deepest biological programming ushers us through every stage of reproduction. Yet even here, the modern world has turned the playing field upside down. The sweetness and passion of sexuality—once a private matter between two people involving innocent responses of soul, heart, and hormones—have been distilled and injected into the ever-present media images that shape our self-concept and our images of love and sex. Our fertility is challenged by the preponderance of artificial hormones in our foods and electromagnetic pollution in the air. The effects of whatever may be unhealthy in our complex lifestyles are magnified during pregnancy. Even our ways of giving birth have been divorced from our natural rhythms and instincts.

Energy medicine has much to offer in each of these arenas. By bringing you into better communication with your energies, your body—the world's foremost expert about your sexuality, fertility, pregnancy, and birthing—is able to talk to you in its own language. Besides helping you re-embrace your natural instincts, energy medicine provides tools for discovering what may have gone awry in any of these areas and supplying effective corrections. These tools can also be used for creating an atmosphere—an energetic environment within your body—that will help you flourish in every aspect of reproduction. This chapter offers basic concepts and methods in each area.

Sexuality

I am part Cherokee. Cherokee grandmothers teach their granddaughters how to pleasure themselves without a man. This is not a self-indulgent practice, but a vital skill for maintaining their independence and self-reliance until they are ready to be with a mate. The original meaning of the word *virgin,* in fact, meant "a woman who owns herself." Healthy sexuality for a woman begins by early on getting to know her body, revering it, becoming adept at pleasuring it, cherishing it as a holy vessel that may one day carry a baby, and developing a fierce intent not to violate it. Owning your body as a young woman also means that you do not casually give away your virginity or your sacred sexual involvement as a barter for being liked, accepted, or popular; as a way of feeding a hunger for love or affirmation; or as a collapse into serving another's desires at the expense of your own integrity. All this is as relevant today as it ever was; I am not suggesting a return to prudish times. I am saying, however, that a young woman may pay a great price for having multiple partners. Aside from sexually transmitted diseases and the decreased self-esteem that is often found in girls who do have multiple sexual partners, the body reacts. The risk of cervical cancer in women who have had multiple partners is vastly increased. Some women who took on multiple partners when they were young lose the capacity to derive deep pleasure from sex even after having found a partner they deeply love. This sad fallout from the exuberance of the free-love era has been reported to me so frequently by my female clients that I am surprised it is not spoken of more openly in these times when we are still coming to terms with the freedoms birth control has opened to us.

THE CHEROKEE GRANDMOTHER'S SELF-EMPOWERMENT SECRET

While I did not grow up in the Cherokee tradition, I was shown the grandmother's self-empowerment secret from an old Cherokee woman when I was in my 30s. I was astonished because I had spontaneously "discovered" it when I was 11, as if it were passed along to me as an echo from my ancestors. Back then, I never thought of this as sexual, nor is it particularly focused on the genitals. Rather it was a sense of immense pleasure and wonderful invigoration. I would not feel free to share it had I not discovered it on my own, but since I did, gather round, ladies.

The Cherokee Grandmother's Self-Empowerment Secret can be approached in two ways, one that is done in total stillness and one that involves movement. The first is the way I discovered it when I was 11. I was lying in the sun on a large boulder behind our house in Ramona, California. It was my favorite place. A deep stillness quieted my body and mind. From this stillness, I began to feel strong pulsations coming up into my body from the earth. It was as if my breath was no longer powered by my own lungs, and instead the earth was breathing me. After a while, the pulsations took the form of a figure-eight energy moving in waves lengthwise through the trunk of my body. Up and over, around and under they went, again and again, in a perfect rhythm. I felt like I was an instrument being played by an exquisite force I couldn't explain. The sensations throughout my body brought a pleasure that satisfied at the deepest level—physically, emotionally, spiritually. It was pure joy, indescribably wonderful to engage these deepest energies that activate sensual pleasure, peace, and restoration.

To this day, when I want to be nurtured and restored, I go into that deep stillness, and the "energy eight" begins to travel. It is not necessary to be on a rock or grass or earth to do this. A bed or floor works just fine. I believe these vital figure-eight energies are always there waiting to be engaged. I just become still, set my intention, and my body seems to do the rest. It does require a degree of surrender. The version that involves movement is a good way to learn the method. My clients have told me that even if it is difficult to attune themselves to these figure-eight energies, or to surrender to them, the physical movements help them experience their latent energies. Eventually the form that does not involve movement may become available to them.

The old woman taught me the physical movement, but then she made a comment that the girls eventually find that they don't even have to move their bodies. They discover that the energy wave begins on its own, just as I had initially experienced it. Here is how she taught it.

GRANDMOTHER'S SELF-EMPOWERMENT SECRET

1. Begin by lying down on your back and take two deep breaths, in through the nose and out through the mouth. You may prefer to close your eyes. Then let your breath become natural and easy. Force and effort do not work with this method. Surrender does the trick.
2. Focus your attention on your root chakra, the lower part of your pelvis. Sense into the energies that are always there. Imagine energy and oxygen moving in and out of every cell.
3. Circle your hips in an undulating up-and-down figure-eight movement.
4. Bend your legs at your knees, and place your hands on your inner thighs, with the sides of your thumbs lying on the creases where your legs and body join (see Figure 5-1). Move your whole body in an undulating figure-eight motion. The liver, spleen, and kidney meridians (all yin, or female, energies) come up the insides of the legs and cross the thighs. The points your hands touch when you take this position can open sexual energy.
5. Allow the energies that will naturally build to flow with these movements.
6. After a time, lay your legs flat, place your hands at the sides of your body, slow your breath, and let go. The sensation of the figure-eight movement may continue on its own and even build.

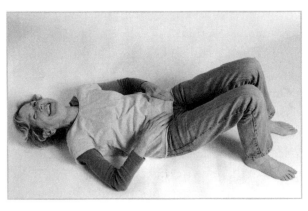

Figure 5-1
GRANDMOTHER'S SELF-EMPOWERMENT SECRET

This obviously is not limited to the Cherokee tradition. It is aligned with the natural architecture of a woman's body. When I shared the method in an energy medicine course for women, a belly dancer in the class said, "That comes out of the Middle East. Traditional belly dancers know this method." Another woman said, "It originated in China. It is a Triple Warmer/Pericardium technique to get heat and pleasure pumping through your body. It even predates Tantric yoga."

EVOLUTION'S JOKE, SOCIETY'S DILEMMA

In our culture, many of us were brought up believing that we should not enjoy our sexuality outside a loving relationship. And there are physiological as well as emotional consequences for women having casual sex with multiple partners. Unfair as it might in some ways seem, our hormones tend to support the traditional model that confines intercourse to a loving, committed relationship. Just to be clear, I am not saying, "Wait for Mr. Right before engaging"—this isn't necessarily even my opinion about what is best for all women at all times—but let's look at the hormonal landscape that underlies such choices.

Your sexual and romantic decisions are based less on clear thinking and free will than on your hormones. It is not that rational thought and well-considered choices cannot counter the urges of our hormones. They can, and they need to. But we should also not underestimate the strong magnetic pull our hormones can exert beneath our radar.

Free will is, in terms of evolution, a recent development. Other species do not enjoy anything on the order of our awareness of choice and freedom to choose. Long before nature began its perilous experiment of granting us free will, however, hormones and brain chemicals were doing the job of ensuring that we would perform the behaviors necessary for survival and procreation. Those programs still supplement and sometimes dominate our choices. The fight-or-flight response to threat is a classic example, as are the urgencies to seek food, water, and shelter. These behaviors are biologically programmed, with biology talking to us through the language of pleasure and pain. Most primal here are behaviors that help us survive. Next are behaviors ensuring that we create the next generation.

In that arena, men and women are wired so differently that we surely must be witnessing nature's sense of humor, penchant for tragic drama, or indifference (though who knows which). Frequent intercourse with multiple partners increases the self-esteem of young men, while withholding intercourse correlates with higher self-esteem in young women.[1] Men's fantasies are more about sex, women's more about romance.[2] Men and women have markedly different biologically driven sexual agendas. Remember the Queen-apparent, the month's single fertile egg, poised to carry the woman's genes forward through impregnation, gestation, and years of nurture and protection. The opportunity might appear some 300 times during the woman's lifetime, with impregnation transforming the drama from a monthly festival into a long-term commitment. For the man to fulfill the minimum requirements in ensuring his contribution to the gene pool, impregnation is the only act in the play that really matters, and he has billions of sperm that are at the ready at any time the opportunity might arise. And his hormones are telling him to seek those opportunities frequently.

Energetically, arousal for a man quickly ignites the genitals, and the energy moves up toward the heart and warms it. For a woman, arousal starts in the mind or heart and then travels to the genitals.[3] The hormones testosterone, vasopressin, and dopamine are also well recognized as driving male sexual pursuits; however, the role of oxytocin is especially entertaining. Oxytocin, the "tend and befriend" hormone, is associated with women bonding with one another and nurturing their families. And indeed, it is often ten times as concentrated in women's bodies as in men's.[4] Except at one particular time—during orgasm. That's when men get to have a deep bonding moment. Their oxytocin jumps to five times normal levels. For a brief instant, nature achieves something close to emotional parity between the sexes. Beyond release, beyond pursuit, attaining this state is part of men's obsession with copulation. The common wisdom that men give love in order to get sex is only half the story—men also want sex in order to feel love. Quickly, however, while the woman is basking in the shared intimacy, the man's oxytocin returns to its normal low levels and she is somewhat deflated when his romantic murmurings are replaced with Monday-night football. While nature hopes that if you can entice him into 20 or 30 of those bonding moments it will begin to involve other parts of his brain and become a habit so he starts to feel love that transcends the exhilaration of the pursuit and stays around to protect your babies, the hormone-driven differences in male and female sexual behavior should not be underestimated. Most specifically, do not assume that the rules generated by his brain are the rules your body should follow.

This is far more than a mere personal matter, and it extends far beyond sexuality. We live in a culture whose governance conforms much more to the influence of testosterone than of oxytocin. When a woman faces stress, her body produces oxytocin, which buffers the fight-or-flight response and causes her to seek ways of connecting with others. After any immediate physical threat, her focus goes to conferring with other women and tending to the welfare of the children. The more she does this, the more oxytocin she produces. The behavior is self-reinforcing. Beyond that, estrogen enhances the bonding effects of oxytocin. Testosterone, on the other hand, which men produce in high levels when stressed, tends to inhibit the bonding effects of oxytocin. When relationships become strained, most women want to talk; most men turn inward and seek solutions in isolation. Ease of communication was not nature's first priority in fashioning our differences. Culturally, we are challenged to advance oxytocin-based strategies to our problems by mediating the fight-or-flight response and elevating "tend and befriend" so it becomes a much stronger possibility for both men and women.

The ongoing suppression of feminine values goes to the heart of whether humanity is going to alter its current rapid course toward extinction, and it impacts our physical health as well as our mental health. While the stifling of the very values that could save us all takes many forms, subtle and blatant, its primal foundation traces to physical domination. The world's tolerance of violence toward women and the resulting suppression of the feminine principle is not just another bothersome problem affecting undeveloped countries. Describing her own waking to the pivotal role of this issue, Sister Mary Eve recounts: "Slowly, it dawned on me that nothing was more important than stopping violence toward women—that the desecration of women indicated the failure of human beings to honor and protect life and that this failing would, if we did not correct it, be the end of us all. I do not think I am being extreme. When you rape, beat, maim, mutilate, burn, bury, and terrorize women, you destroy the essential life energy on the planet. You force what is meant to be open, trusting, nurturing, creative, and alive to be bent, infertile, and broken."[5]

Why is this detour into social commentary relevant for a health book? Because if you are dancing to a drummer that is seriously out of phase with evolution's best strategies, as coded in your hormones, your body pays a price. Most specifically, your energies become conflicted within themselves, and your health, your happiness, and your sexuality may all be compromised. When clients lie down on my table, not only do their energies reflect their genetic predispositions and physical circumstances, they also reflect their life choices. This is just as true for men as it is for women, and nature's trick of providing us with such different hormone-based strategies for bonding, resolving problems, and reproducing is the fundamental puzzle whose solution our collective survival depends on.

OPENING SEXUALLY
IS MORE THAN PUSHING POINTS

Energy medicine shows you which points to hold, massage, or tap in order to awaken your sexual and other energies. More than holding points, however, the first order of business is to attend to yourself as an energetic being. My very first client was a well-known personality who was in his late 80s. I hadn't actually set up a practice yet, but I had just taught my first energy medicine class, and one of the participants told him about me. While he had a number of health problems, he had only one goal in seeing me. He had lost the ability to have an erection. He wanted a girlfriend but didn't feel it was fair to a woman to pursue her if he couldn't perform sexually. This was definitely not my area of expertise. But I did believe that if I got his body's energies flowing, it could only help. So that is what I set about doing. Wherever his energies were blocked, I opened the channels.

While every health problem in his body showed improvement within just a few sessions, there was only one he wished to focus upon. A successful man, he was blunt and direct about his priorities. I received a triumphant phone call from him one morning announcing that he was holding it and it was hard. He told me I was worth my weight in gold and that he was going to get himself a girlfriend. He did. That was a very instructive experience for me. I didn't know the points that are used for enhancing sexual performance. I didn't know cures for anything. I only knew how to balance the energies in the body.

So all of the methods in this book for balancing your energies contribute to your becoming a more robust sexual being. However, a few techniques can open sexual energy that has been chronically blocked or inhibited.

Opening Your Sexual Energies

Aside from childbirth, sex is the ultimate physically intimate act, usually more so for a woman than for a man because she takes her partner *into* her body. It is far beyond our scope here to address all the emotional, interpersonal, and spiritual dimensions of your sexuality, but I can share some techniques for sparking the energies that affect it.

Getting Your Sexual Feelings to Flow. In energy medicine, the meridians most involved in healthy sexuality are the kidney, spleen, and liver meridians.

- The kidney meridian, in some traditions, is considered the storage area of sexual energy, and while I don't quite see it that way, it is a strong force in keeping our sexual energy flowing.
- The spleen meridian bolsters the kidney meridian in this function, and also supports all the organs involved in female sexuality.
- The liver meridian is able to open blocked energies, and supports the muscles and ligaments that are part of sexuality.

To maintain a healthy sexuality, or to open blocked sexual energy, keep your kidney, spleen, and liver meridians humming. Use any combination of the following methods:

a. In the Cherokee Grandmother's Self-Empowerment Secret, placing your hands on the insides of your thighs as your body undulates stimulates all three of these meridians. For some women, simply keeping their hands there gets energy flowing. While this can initiate a warm, liquid-like feeling and you may not want to be any more intrusive than that, gently massaging the points on the insides of your thighs in a circular motion will also stimulate these energies.

b. You can use your hands to keep the kidney, spleen, and liver meridians open and flowing. Your hands are electromagnetic, and moving them along segments of the meridian lines can invigorate these meridians. Place your open hands at the insides of your feet. With your hands spread, draw them very slowly and deliberately up the insides of your legs to the tops of your inner thighs.

c. Acupressure points for the kidney, spleen, and liver meridians are also situated on the creases where your legs and body join. Gently press or massage along the creases to stimulate the energies in these meridians.

d. Massaging any tender points on your pelvic bone also keeps the pathways clear so sexual energy will be able to flow.

Deliberately propelling the kidney, spleen, and liver energies through your thighs once or twice each day changes the energetic habits in your entire pelvic area so that sensuality and sexual excitement are more readily accessible. Regularly pressing the points on your inner thighs, leg-body creases, and pubic bone also has the effect of keeping your consciousness attuned to yourself as a sexual being.

The Back Side of Sexuality. The energies held in your buttocks are ruled by the circulation-sex meridian. If these energies become stagnant or blocked (descriptions of personality characterizations like "tight-assed" often have a physiological counterpart), your capacity for sexual enjoyment is inhibited. You want to release these energies so they can flow forward and become fully engaged during sex. One way to know if your circulation-sex energies are blocked is by pushing in on your buttocks and seeing if there are areas that are sore. Sedating the circulation-sex meridian frees congested energies and helps keep the energies moving throughout your root chakra, an energy center that is strongly involved in your sexuality. Sedating the circulation-sex meridian (see page 148) not only opens a door to sexuality, it can also help stabilize an irregular menstrual cycle.

Figure 5-2
CENTRAL SEXUAL CHANNEL
(HAND POSITIONS)

Opening the Central Sexual Channel. Another set of points, which affect the central and governing meridians and a radiant circuit called the penetrating flow, open the energies involved in deep, whole-body sexual experiences.

a. Lying down, cup your hands and bring the three middle fingers of each hand together, back to back, fingernails touching. Place them about an inch above your pelvic bone (see Figure 5-2).

b. Push your fingers in. Lie this way quietly for about three minutes. Notice the movement of energy.

The Vortex Revival. There is an energetic parallel to the way a woman's genitalia differ from those of a man. At the root or sexual chakra, the energies pull in, like a magnetic vortex, in a woman, while in a man there is a spiraling force that pushes outward. One receives, one projects. It is because energies spiral into our bodies that we, as women, feel so deeply. Men, who express pride in being more objective than us, are processing their emotions in the spiraling chakra energies *outside* their bodies. To play off lyrics by Janis Joplin, "Objective's just another word for nothing gets inside." Of course all men and all women have both yin and yang energies in varying combinations, so the generalizations do not always fit and are not really fair, but for the most part, for women more than for men, our relationships penetrate us, not just physically but energetically. We take our partners into our bodies physically, emotionally, and spiritually, and if we are hurt too much or too often, that spiraling inward force, which draws a partner's energies into our own, begins to diminish and lose its magnetic pull. Whether you have been wounded or just want to give that magnetic pull a boost, the following exercise energizes and revives the natural vortex energies in your root chakra.

The Vortex Revival Preliminaries
(time—about 90 seconds)

1. Sitting in a chair, place your hands on the top of your hip bones and slowly and firmly move them down the fronts and sides of your legs and massage the energies off your feet, like squeezing a tube of toothpaste. Then slowly and firmly bring your hands back up on the insides of your legs and out to your hips. Do this two or three times.

2. On a deep in-breath, draw your hands straight up the middle of your body (along the central meridian and over the penetrating flow) from your pubic bone to your chin. Release your breath.

3. On another deep in-breath, push your fingers up over your chin and cheekbones with

Figure 5-3
THE VORTEX REVIVAL
PRELIMINARIES

some pressure until you are cradling your face in your hands, with the middle fingers at your temples (see Figure 5-3). Release your breath.

4. Pushing your fingers up from your chin and onto your face opens the flow of the stomach, small intestine, large intestine, gallbladder, and triple warmer meridians.

5. Do a Triple Warmer Smoothie (page 107) and a Hook-up (page 62).

This in itself opens the meridians involved with sexuality, and you can stop here. But it is also a great preliminary to the Vortex Revival, which is more like a meditation.

The Vortex Revival (time—five to ten minutes)

1. Lie down on a bed or on the floor with your palms up and your hands open.
2. Imagine a swirling, spiraling energy at the end of your feet.
3. Imagine it now very slowly and deliberately being drawn into your body, coming in through your feet. Energy follows imagination!

Figure 5-4
THE VORTEX REVIVAL
(HAND POSITIONS)

4. Imagine it coming to your pelvis and climbing up through the center of your body so there is a swirling energy within and above your pelvic area.

5. Imagine this as a spiraling force that literally sucks energy from outside you into your body. This is a vortex revival moment! Spend some time playing with these energies.

6. Linger here, or take it further if you like, by experimenting, in your imagination, and expanding the size of the swirls until, with the spiral centered at your pelvis, it grows to encompass your heart. This connects the heart and genitals in a full-body experience.

7. End by placing one hand on your second chakra (just below your navel) and the other

on your heart chakra. Within one to three minutes, you are likely to feel a strong connection between these chakras (see Figure 5-4).

Sometimes in my practice, a woman wails and sobs just from feeling the energies moving again. This energy is very real, and it originates within you, far different from the physical connection with a partner. A variation of the Vortex Revival with a partner or even good friend can, however, be a wonderful experience.

The Vortex Revival with a Partner (time—five to ten minutes)

1. As you lie on your back, your partner puts both hands on your hips and gently, slowly, firmly, pulls her/his hands down the outsides of your legs.
2. Then with more pressure, have your partner pull the energy off each foot, with both hands on one foot and then the other.
3. When you do the Vortex Revival on yourself, you use your imagination to engage the spiraling energies. When you have a partner, the electromagnetic energies of his or her hands can interact with your energies to make the exercise even more powerful. Your partner sits or stands about 24 inches away from the base of your feet and, with a flat open hand, begins to make slow, counterclockwise circles in the energy field of your feet (see Figure 5-5).
4. If he or she moves slowly enough, you will begin to feel a spiraling force at your feet that begins to move up your legs toward your pelvis. Your partner can support this movement by coming closer and closer to your feet with the spiraling motion.
5. Focus on the way this spiraling energy engages your pelvis or even moves up to your heart.

Figure 5-5
THE VORTEX REVIVAL
WITH A PARTNER

171

6. You may experiment with using your imagination to expand or intensify the spiral.

7. End by placing one hand on your second chakra (just below your navel) and the other on your heart chakra (see Figure 5-4). Within one to three minutes, you are likely to feel a strong connection between these chakras. Alternatively, your partner's hands can be placed over these chakras.

ENERGY TECHNIQUES FOR ENHANCING SEXUALITY WITH A PARTNER

Playfully thinking outside the box does not require techniques. You follow the energy and the energy follows you. Start with creative kissing. Here are Three Great Kisses:

1. Slowly, softly, mindfully, let your lips linger behind one of your partner's knees. The neurovascular points that sit there are some of the sexiest points you can stimulate, and just before the tickle zone is an exhilarating space.

2. One of the most romantic spots for a kiss is on the top of the upper eyelid. Gently and fully attuned to the connection you are making, soften your lips and know you are stimulating the meridian that helps two people cozy up with each other.

3. A long held kiss on the side of the cheek stimulates several energy points that help awaken the flow of love.

The most frequently reported reason couples who have been together for a long time seek sex therapy is because one or both have lost interest in their sexual relationship, or as the professionals label it when they bill insurance companies, the patient has an "Inhibited Desire Disorder." While it is, of course, important to work with communication, resentments, and failed expectations, a behavioral approach is often what is most effective in successfully helping a loving couple respark their sexual desire for each other. Even as simple a technique as the Three Great Kisses can get the energy into a habit of flowing between the two partners.

Margot Anand's classic *The Art of Sexual Ecstasy*[6] is a wonderful guide to heightening the experience of making love, grounded in knowledge of ancient wisdom such as Tantric[7] and kundalini yoga, as well as insights from contemporary sex therapy. Making love can be elevated into a true art form, yet at its core is energy. Making love can become a shared journey in discovering the magic of the body's energies and how the energies of two bodies can interact, blend, dance, and merge. It is an epic adventure. By working with your body's energies and flows, you will already be enhancing your sexuality, but it is also possible to focus energy techniques specifically for sexual enhancement. This, however, like many of our topics, could be the basis of another book. Fortunately, Anand's book is very attuned to the dimension of energy within sexuality, so I'm taking the afternoon off.

HEALING WOUNDS/FEELING SAFE

When we are young, the passion of the moment may fuel great sex. If we are innocent and undamaged, great sex opens our souls to our partners. But let's face it, in most of the scenarios likely to follow when we are young, we get wounded. Not fair to us. Not fair to the partner, who was just looking for an entertaining way to spend the evening and is genuinely surprised when we start to stalk him. Okay, gross exaggeration, but the hormonal aftermath of great sex may be very different for men than for women. For us, great sex *is* risky business. In the wild, if you become pregnant and he leaves you, your literal survival is threatened. But even in a world where we can control pregnancy and have many levels of social support to fall back on, abandonment after great sex is still a primal loss and our wound may be soul deep. Both sexuality and survival depend on the energies of the root chakra, and therein lies the rub. Your root chakra doesn't register the subtleties. If such abandonments occur a number of times, or even if you have one whopper, the root chakra energy may shut down your sexual juices. The passion of the moment may become very difficult to enter.

As we mature, great sex grows from having confidence in who we are and in the truth of our own experience. If you are guided primarily by a focus on your partner's needs, desires, and judgments about you, your sense of self, and soon your enjoyment, get lost in your sweetness. Sustained great sex also grows from having confidence in your partner's love and caring for you. Feeling victim to your mate's expectations isn't particularly sexy; succumbing to pressure closes down your pleasure circuits. Establishing a deep foundation of trust leads to emotional safety and the intimacy that allows you to laugh, or cry, from the depths of your soul. It is a romantic container that is fashioned both within and beyond the bedroom.

If we have been deeply wounded, sometimes the wound does not even fully reveal itself until we are graced by profound love. Women who have been sexually or physically abused may not be able to open fully to their own sexual pleasure without revisiting the traumatic past and healing emotional wounds that may trace back to childhood. This sometimes may require a substantial course of psychotherapy, but energy therapy is proving itself to be able to markedly speed the healing process. Sometimes if you can heal the emotional wound at an energetic level, the psychological healing follows rapidly. The approach, applying energy techniques for healing emotional wounds, is called energy psychology. While it is beyond the scope of this book to go into that complex territory, here is a case history that is typical of what can be rapidly accomplished. It is from the book on energy psychology that David and I wrote with Gary Craig.[8]

Sandy and her partner came to one of our colleagues[9] for premarital counseling. Among the issues they were concerned about was their sexual relationship. Although Sandy had been married before, she found herself reacting with uncontrollable negative feelings when her fiancé initiated sexual play. He was willing to be patient, kind, and understanding, and he seemed genuinely interested that sex be a shared experience. While she freely acknowledged that she had no problems with his attitude, she still would usually became upset and turned off by his overtures. They asked for help with this problem, and a private session with Sandy was arranged.

When she came in, the therapist gently asked, "Is there something in your earlier years that you could talk about?" She immediately burst into tears. Red blotches appeared on her skin, and her words were punctuated with heavy sobbing and gasping as she began to relate her story: "When I was seven years old, we lived in [a small rural town]. One day my stepfather took me for a walk down a country road. It was in the summer. We hiked up the side of a hill. Then we stopped. Then he took off all my clothes. Then he took off all his clothes."

At this point she was scarcely able to breathe. The therapist stopped her and said that it was not necessary to go any further. He had her state her distress rating about the memory on a 0-to-10 scale. Not surprisingly, it was a 10. He then led her through a tapping sequence similar to the Meridian Energy Tap [page 80]. Her intensity dropped from 10 to 6. After another round of tapping and the use of affirmations such as "Even though I still have some of this upset, I deeply love and accept myself," the intensity fell to 2. Then another round of tapping was completed, along with another simple affirmation.

By this time, Sandy was breathing quietly. Her skin was free of blotches, her eyes were clear, and she was looking at her hands, lying folded in her lap. The therapist said, "Sandy, as you sit there now, think back to that hot summer day when your stepfather took you for that walk down that country road. Think about how you hiked up the side of that hill until you stopped. Think about how he took off all your clothes. Think of how he took off all his clothes. Now, what do you get?"

She sat there without moving for maybe five seconds, then looked up calmly and said, without strong emotion, "Well, I still hate him." The therapist, after agreeing that hating him seemed a perfectly human response, then asked, "But what about the distress you were feeling?"

Again she paused before answering. This time she laughed as she said, "I don't know. I just can't get there. Well, that was 20 years ago. I was just a little girl. I couldn't protect myself then the way I can now. What's the point in getting upset about something like that? . . . I never let that man touch me again, and my kids have never been allowed to be near him. I don't know, it just doesn't seem to bother me like it did."

After this single session, Sandy no longer experienced negative feelings in response to her partner's sexual advances. In a two-year follow-up, she reported that the problem was "good and gone," and her partner, now her husband, confirmed that there was no sign of the former difficulties. Such shifts in relationship to a traumatic memory that has been emotionally cleared using an energy intervention are surprisingly common. While therapy around past abuse may require skilled assistance, the Meridian Energy Tap (page 80) can be adapted and helpful in many contexts, and it is quite harmless. If you combine the tapping with a memory or a worry that plagues you, doing the tapping sequence again and again, the memory or worry will often lose its grip.

Fertility

You are designed to have an egg poised for fertilization every month. You are designed to have hormones raging through your bloodstream that cause your brain chemistry to urge you to find a mate. You are designed to become pregnant, carry, and birth a baby, and to quickly become pregnant again and again and again, from your early teens into your forties, or until your body says "enough" and you die in childbirth. Meanwhile, nearly every human culture has been trying to find ways to put the brakes on nature's prolific plan for you. Sometimes the brakes, however, are permanently applied, through culture or biology, in a woman who desperately wants to conceive a baby.

At least 100 women or couples have come to me over the past 30 years asking for help after having difficulty conceiving. All but three of them became pregnant. Usually if there was serious and obvious physical damage interfering with the ability to become pregnant, I would not be consulted, which admittedly did up my percentages. But unless there is a decisive physical obstacle, conception is usually possible. There are many factors involved in infertility, and the circumstances of the three women who did not conceive are instructive. One of them eventually discovered that her fallopian tubes were totally closed off, although this had not been detected in an earlier medical exam. Another, after substantial individual work with me, learned that her apparently virile young husband had an extremely low sperm count. A third couple was already headed toward divorce and hoping at some level that having a baby would fix their marital problems. The sad truth was that their energies did not easily dance together, and the tension between them worsened their incompatibilities. While energetic differences form the spark of romance, the topic of another program,[10] stress and personality incompatibilities exaggerate the energetic differences at all levels. But for everyone else, baby pictures were not far off.

Pregnancy occurs when a woman releases an egg from her ovary, it travels into the fallopian tube, it is fertilized by a sperm, and it continues down the tube to the uterus, where it implants itself and begins to develop. Infertility means that this sequence is not able to be completed. Medical exams and laboratory tests can often determine the physical reasons for infertility. Among the most common are blockages in the fallopian tubes or damage to the ovaries. Pelvic inflammatory disease, often caused by a sexually transmitted infection, may damage the fallopian tubes. Endometriosis, where tissue from the uterine lining grows outside the uterus, may also damage the fallopian tubes as well as the ovaries. Fibroids may crowd the uterus so that an egg cannot implant or a baby is unable to grow. The sperm or egg may be damaged, sometimes from environmental causes such as radiation or pollution or from the use of tobacco or other drugs or simply from age. Attempts to intervene have led to the growth of a $2 billion fertility industry. But, as an article in *Time* points out, "the more doctors have to intervene with drugs, needles and surgery to get sperm to meet egg, the greater the chance that something will go wrong."[11] In addition, according to Randine Lewis in her excellent book *The Infertility Cure*, hormonal imbalances "contribute to 40 percent of infertility, yet are considered untreatable by conventional Western Medicine."[12] A woman's body can, however, in my experience, be brought into a hormonal balance that supports pregnancy by using energy medicine, with larger amounts of the hormones needed for conception being produced naturally.

Energy work can change chemistry, open pathways, get the body to expel toxins, enhance the health of the blood, and help the body cycle regularly. It can keep the uterus flushed and healthy and even repair it. And of course, it can change energy habits (an energy habit might, for instance, promote the growth of fibroids or keep your cycle at 23 days instead of 28). It is actually surprising how many ways energy medicine can help a woman become fertile.

Even physical obstructions were often energy obstructions first. By correcting imbalances in the energy field, physical blockages can literally be dissolved. Dietary changes are a natural way to shift the body's energies, and they can be used to increase the chances of conceiving. Commonsense notions that it is probably not a good idea to mess too much with the way Mother Nature produces our food are supported by a study of the impact diet has on fertility. The findings, based on the experiences of 18,000 women who wanted to conceive, were striking and surprising. Whole milk and rich ice cream promoted pregnancy, while skim milk and other "light" dairy products such as low-fat yogurt actually worked against fertility. Natural, unsaturated fats increased the chances of pregnancy, turning on genes that enhanced fertility, while trans fats had a detrimental effect. A book by researchers from Harvard Medical School, *The Fertility Diet*, presents practical guidelines based on the study.[13]

An "energy habit" that often impacts fertility is what I might call "chronic fatigue of the reproductive organs." Everything seems slowed, from the movement of the egg to the delivery system for the sperm. I suspect that this is a factor in those infertility cases where all the standard tests show up as normal. Stagnation in the liver meridian or diminished levels in the energies affecting the thyroid can also interfere with the body's ability to conceive, as can a dominance of a woman's yang energies (hot, bright, quick, aggressive) over her yin energies (cool, dark, slow, receptive). Lewis calls conception "a fragile miracle that can be affected by any one of a thousand factors."[14] By counteracting energy habits that interfere with conception and bringing the body's energies back into their natural balance, the body's remembered wholeness and script for making a baby can be restored.

Emotions—fears or ambivalence or interpersonal difficulties with one's partner—can also cause energy patterns that interfere with pregnancy. There is a saying in traditional Chinese medicine that "fear scatters *chi*," and in my experience, it is indeed harder to retain the embryo in the uterus when anxiety and worry are too strong. Confronting emotional concerns energetically, using techniques from energy psychology, can remove some of their charge, making it easier to directly and consciously address their causes.

AN ENERGY APPROACH TO FERTILITY

A woman's body can be gently nourished with energy methods, like tending a tree in an orchard, and encouraged to bear fruit. Many women, however, crumble under the sorrow and fear of believing that they are unable to have a child. It has been a remarkable privilege to first demonstrate, through energy tests, that the door is not closed, and then to help these women systematically shift the energies that are interfering with pregnancy, and to finally hear the triumphant announcement that they have conceived.

When infertility is an issue, unless it involves collapsed tubes or other substantial physical damage, the first steps are to regularly do the Five-Minute Daily Energy Routine (page 51) and to be sure your energies are not running in a homolateral pattern (page 67). With infertility, spleen meridian is also almost always involved. Spleen meridian is considered the "mother" of all the body's energies. It tries to nurture you whenever and wherever there is threat. It will create inflammation if you get a splinter in an attempt to remove the invader and initiate a healing process for the tissue. Its primary strategy of protecting you, however, is to mobilize your entire body to stay strong. In the sense that a lion will attack the weakest wildebeest in the pack, and the weakest individuals, constitutionally, are the most vulnerable to invasion by harmful microorganisms, a strong energy system is one of the best ways to protect yourself from outside invaders or internal breakdown. Spleen meridian works to keep your whole system strong. It is a community builder, orchestrating your organs, blood, cardiovascular, and other systems.

Triple warmer, on the other hand, which also protects you in the face of threat, tends to use attack as its primary strategy. If spleen mobilizes your "inner mom," triple warmer meridian mobilizes your "inner militia." In our culture, just as patriarchal values and militaristic strategies dominate in the outer world, triple warmer tends to dominate in our inner world. In the balance between triple warmer and spleen, triple warmer can conscript energy from spleen to fight its battles. This diminishes your overall strength, vitality, and joy. Triple warmer perceives constant threat in the modern world, with its pollutants, stresses, and chemicals that were not there while it was evolving. So for many people, triple warmer is in perpetual reactivity and overcharge, while spleen meridian is in perpetual depletion. Among the many problems this causes, if spleen meridian is depleted, less support is available—in keeping with its nature as the "mother" energy—for becoming pregnant or for carrying a baby.

Spleen meridian, in fact, governs the sea of blood and the cycle of menstruation. It governs whether the blood expels completely or coagulates and stays in the body, where it can grow as endometriosis or fibroids. It governs metabolism as well as how quickly the body can restore and replace natural hormones. It governs the body's ability to renew itself. And it governs the vitality of the sexual organs and the uterus, keeping them in a state that is inviting for a baby. Strengthening spleen meridian is often the first order of business when working with infertility. It is also a way of keeping your entire body strong.

To nurture spleen meridian, you can strengthen it directly, but it is often more effective to calm triple warmer so that it is not perpetually pulling energy from spleen meridian. An overactive triple warmer can, in fact, undo your efforts to strengthen spleen directly. I personally had an extreme imbalance between triple warmer and spleen meridian that was the underlying energetic cause of an assortment of medical conditions, though they seemed unrelated. Strengthening spleen meridian and calming triple warmer began to restore my health. The epidemic imbalance between triple warmer and spleen creates energy patterns that are dysfunctional, and sometimes you need to be diligent to shift these habits. You've already been introduced to each of the techniques in the following sequence. Using it two or three times each day will do a great deal to keep your spleen meridian strong:

Spleen/Triple Warmer Balancing Sequence
(time—about 3 minutes)

1. The Triple Warmer Smoothie (time—20 seconds, see page 107).
2. The Triple Warmer Tap (time—about a minute, see page 108).
3. Spleen Meridian Flush and Tap (time—less than a minute, see page 136).
4. Neurovascular Hold (time—1 to 3 minutes, see page 151).
5. Triple Warmer/Spleen Hug (page 110)
   Each of these supports a strong working partnership between your spleen and triple warmer meridians. The "Hug" is particularly convenient because you can do it while you are talking with someone or watching television, and many other times throughout the day. It is something you don't have to think about.

Fertility Stimulation Points (time—less than 2 minutes)

In addition to the vital role of spleen meridian, attention to your kidney, liver, and circulation-sex meridians also supports fertility. Stimulating the following sequence of points will help optimize all three meridians.

1. Use two or three fingers to tap the points directly in front of the inside anklebones for three deep breaths. You can tap both sides simultaneously (Liver 4).

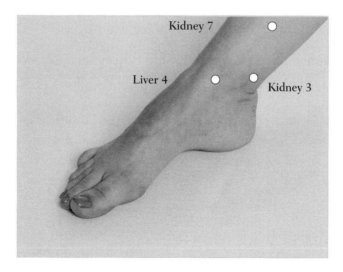

Figure 5-6
FERTILITY STIMULATION POINTS

2. Tap in back of your inside anklebones, again on both sides for three deep breaths (Kidney 3).
3. Tap two thumb widths above the inner anklebones for about three deep breaths (Kidney 7) [see Figure 5-6 for all three points].
4. Push your breasts up and find the points directly behind your nipples. Massage as firmly as you can tolerate for several seconds (circulation-sex neurolymphatic reflex points).
5. Do the Triple Axis Hold from the PMS chapter (page 134) to stimulate the endocrine glands related to fertility.

One additional concern, even after doing all of the above, is that if the energy in the circulation-sex meridian is not flowing well, it gets trapped in the gluteal muscles, interfering with all the organs and processes related to conception. You can know if this is occurring for you by simply pushing in with your fingers or thumb on your buttocks to see if there is tenderness. If there is, you can "loosen up" the circulation-sex meridian by sedating it (page 148). You will probably find that there is less tenderness in your glutes immediately after this procedure.

PHYSICAL CONDITIONS THAT MAY PREVENT CONCEPTION.

Four physical conditions that may prevent conception are problems with the ovaries, fibroids, endometriosis, and the impact of age on fertility.

Energies Not Flowing through the Ovaries. Your ovaries are situated directly on the stomach meridian line. If energies are not flowing well through the ovaries, you are less likely to become pregnant. This flow can be enhanced by sedating the stomach meridian (page 143), followed by tapping the points at the top of your cheekbones, directly beneath your eyes (Stomach 1 and 2).

Fibroids. Fibroids are noncancerous tumors composed of blood and disorganized tissue. If blood doesn't completely leave the uterus, it can coagulate, adhere to the uterus, and grow. You are wise to take measures to prevent this. As a fibroid grows in the uterus, in a tragic case of mistaken identity, the body begins to fiercely protect the fibroid like a baby and to marshal all available resources to help it grow. One of the simplest ways to keep the blood from staying trapped in the uterus is to place a heating pad over your uterus for 10 to 15 minutes on the days your period is winding down and on the day that follows. The heat will facilitate release of the menstrual blood. The same meridians that are involved with fertility—the spleen, triple warmer, kidney, liver, and circulation-sex meridians—may contribute to fibroid growth if they are compromised. So each of the procedures mentioned earlier for reversing infertility also addresses fibroid growth. Particularly important is the liver, which processes hormones. Sedating the liver meridian (page 146) restores balance among your hormones and aids in the prevention and shrinking of fibroids.

Dietary supplements can also be helpful. Essential fatty acids such as those found in flaxseed, fish oils, and evening primrose oil may help prevent and dissolve fibroids, as can nettles, kelp, motherwort, and a combination of Chinese herbs called the "blood movers," including cinnamon twig, peach kernel, and red peony.[15] Fibroids are discussed later in the chapter in the personal story about my daughter: "Energy Medicine and Conventional Medicine Together in Concert."

Endometriosis. The endometrium—the mucous membrane that lines the uterus—can grow outside the uterus, with its cells somehow migrating and becoming embedded in sites such as the cervix, ovaries, fallopian tubes, vagina, colon, rectum, or bladder wall. Misplaced endometrial cells have also been found in the abdominal wall muscles, lungs, nose, and even the brain.[16] Wherever they are, however, they respond to fluctuations in estrogen and progesterone just like the endometrium, so they cause bleeding during menstruation. This blood often then becomes trapped. Severe inflammation, scarring, terrible cramps, painful intercourse, and infertility may result. Energetically, the same problems with the spleen, triple warmer, kidney, liver, and circulation-sex meridians that play into infertility and fibroids may also be the cause of or a large contributor to endometriosis. Both fibroids and endometriosis trace to blood that gets trapped and stagnates. The challenge is that since menstruation does not clean out the sites where the stray endometrial cells have lodged, the body must be helped to clear them out. Supplements that can assist include flaxseed oil, evening primrose oil, Pycnogenol, and anti-inflammatory enzymes such as bromelain. The energy medicine approach to preventing and reducing endometriosis, beyond a Daily Energy Routine, involves restoring balance to triple warmer (page 110), strengthening the spleen meridian (page 132), flushing the stomach meridian (page 138), massaging the liver meridian points on the hands and feet (page 47), and sedating the kidney (page 145) and circulation-sex meridians (page 148).

The Impact of Age on Fertility. Of the women who have had an impulse to kill me, up at the top of the list were those who came for help as they entered menopause and wound up getting pregnant. I was balancing their hormones and energies so they would feel better as they went through menopause but, in regulating these energies, they became fertile again, sometimes after years of sexual activity without birth control. More than the other meridians, the spleen and kidney meridians tend to weaken with age. Since meridians are energies and not organs, however, it is possible to keep these important energies youthful and vital. The Daily Energy Routine, as well as the procedures listed earlier in this chapter, is valuable, not just for fertility, but for, as the saying goes, dying young as late as possible.

Wherever the hands on your biological clock, fertility depends on healthy eggs. If you are racing against time, you can extend your fertility finish line by keeping yourself youthful and vital through keeping the spleen and kidney meridians, in particular, strong and balanced. Healthy body, healthy eggs. While many herbal remedies may also be helpful, you might consider in particular antioxidants, such as coenzyme Q10, which eliminate the free radicals that inhibit healthy cell duplication as we grow older.

Despite all the ways it is possible to increase the likelihood of having a child, using both energy medicine and modern technology, motherhood simply is not the fate of every woman. Many of my closest friends were unable to have or chose not to have children. But they bring a quality of incredible love to my girls, to other people's children, and to the world. I am often humbled by the love and wisdom of these "mothers of the heart." As Randine Lewis says in the closing of her powerful treatise on overcoming infertility: "You never know what place you will fill in the universal plan, but I do believe with all my heart that the love that makes us want to be parents is not meant to go to waste. . . . If there is a divine plan and we are placed on this earth to learn and grow, then perhaps the lessons of our souls are taught through those who are put and who are not put in our lives. Ultimately, however, we must recognize [that] the children we wanted so much and have done so much to bear are not really ours to begin with."[17]

ENERGY MEDICINE AND CONVENTIONAL MEDICINE TOGETHER IN CONCERT: A PERSONAL STORY

My daughter Tanya began feeling pain in the area over her ovaries when she was in her midthirties, and the early stages of fibroid growth were detected. Her symptoms included long periods with heavy bleeding, painful intercourse, bladder incontinence, blurry vision, and pain in her pelvis, back, legs, and arms. She also had many symptoms suggesting hormonal imbalances, including extreme PMS, terrible headaches, depression, anxiety, exhaustion, weight gain, dry skin, and hair loss. Poor baby! She was living in Boulder, and my work had me traveling almost constantly. When I was able to visit and work with her, it was obvious which energies were most problematic: the second chakra, the stomach meridian (which travels right over the ovaries), and the small intestine meridian (which governs the abdominals). The spleen/triple warmer balance is also important in almost any illness, with both also playing a significant role in hormone regulation. Each also needed attention. I'd get these meridians back into a healthy flow, Tanya's pain would be relieved almost instantly, her hormone balance would seem restored, and

I'd give her homework to foster further healing. I'd leave for another three months, feeling quite sure the fibroid would now shrink and the pain would not return, as this had always been my experience when I treated clients who presented with similar symptoms.

Following each visit, I was surprised to learn that Tanya's pain had returned. It was clear to me that the recurrent pain, headaches, and the fact that the fibroids were not shrinking were all related to a hormonal imbalance, but it was not clear why the energy treatments were not reversing that imbalance. I was again learning what I'd learned many times before, which is that when I am close to someone, I can't be objective. And it's not just about wanting to help too much or having too much stake in the outcome. It is a literal change in how I perceive the energies. The best way I can explain it is that I couldn't tell where I ended and she began. This was very confusing because I usually know exactly whose energies I am looking at when I am interacting with a client. I was traveling around the world being represented as this great healer, yet I wasn't having success helping my nearest and dearest with a condition that had been relatively easy to turn around with many others.

Ever my daughter, Tanya had been trying to manage her condition exclusively with holistic remedies. Some of the best complementary health practitioners in Boulder had worked on her, but even they could not crack the code. She felt much better after energy sessions. They always helped. But the benefits did not last. In retrospect, taking medical tests that would have resulted in her being given bio-identical hormones would have served her better.

Even with all the care she was receiving, the fibroids did not shrink. In fact, after years of not increasing in size, they grew wildly during a nine-month period following an emotionally significant event that underscored the fact that she was now forty years old, had not had a child, yet desperately wanted one. It was, in fact, as if she were pregnant. All the healers who worked with her during this period commented that it was like a baby was inside her and her body was protecting that baby. The feeling they all reported was that it was as if the fibroid was defiant. No interventions could slow its growth.

She now looked pregnant, and many other symptoms began to appear. Because so much blood was feeding the fibroid, she became anemic and extremely weak. She had difficulty lifting anything, and had to literally climb on all fours to get up the stairs in her home. Her eyesight became bad. She was having heavy periods with blood clots. Her doctors at Kaiser recommended a blood transfusion and a hysterectomy, but as her medical problems become more apparent, Kaiser (coincidentally?) dropped her insurance based on an administrative technicality. This turned out to be a fortunate injustice. She still wanted to have a baby, so she fiercely resisted the recommendation of a hysterectomy, and a deep instinct told her not to have a blood transfusion despite strong medical advice that her life was in danger if she didn't. We found an herbal combination, a powerful iron supplement called Floradix. It went into her cells as if it were being injected, and she brought her extreme anemia to a normal state. Her doctor expressed amazement.

But even as the Floradix was helping her anemia, and the energy workers were making her feel better, her fibroid continued to grow. Strangers would make conversations about her baby, and airport security would direct her away from X-ray devices. We were all in denial, hoping for that magical natural cure that would shrink the fibroid without her having to undergo a hysterectomy.

The final wake-up call was during a long-planned trip to Hawaii. After landing, Tanya noticed a hard swelling near her left ankle. That leg was also in terrible pain. She called me, and I offered some suggestions to her and her partner, including energy exercises to reduce the swelling and help with the pain. These did give some relief, but the pain continued over the next three days, when they called again. Not only was she still battling the pain, the lump she had noticed on her ankle was now traveling up her leg through a vein she could see on the outside of her leg. It was above her knee when she called on the third day. I was alarmed. I told her that it sounded like a blood clot, a thrombosis, and that she had to go right away to a doctor to find out. It was evening, and she was not sure where she could go. We decided that in the meantime, I would do a long-distance healing. Tanya lay down to receive it. I realized after I got off the phone with her that I didn't know to which leg I should direct the healing. But the moment I began to attune to her, the inside of my left leg began to hurt terribly. It turned red and began to swell. There was no doubt now which leg it was. I continued to send the healing energy to her until my own pain diminished and the swelling and redness in my leg had disappeared. I knew something major had been accomplished. She called me right afterward and said that the swelling had gone way down and the pain was gone. Meanwhile, her partner had been on the phone, and he had located a walk-in clinic. They went and the doctor immediately identified it as a blood clot, which she confirmed with an X-ray. She was very interested to hear from Tanya how the energy work and distant healing had apparently caused the clot to shrink and stopped the pain. When Tanya was not enthused about the medication being recommended, the doctor said, "Okay, continue what you're doing and see if you can make it go away." But she warned that Tanya still wasn't out of danger. She said it would be life-threatening if it reached her groin. Tanya also would not be able to get on the plane to return home unless the clot was gone completely. By the time of Tanya's planned return, the deep thrombosis had become a superficial thrombosis and the doctor assured her that she was out of harm's way.

With this close call behind us, we realized that we needed to consider having her fibroid surgically removed, even though the idea terrified me. Besides my bias that many doctors routinely put people at far more risk than is necessary, both my parents had died of medical mistakes made in hospitals. Fibroid tumors affect at least half of all women. The primary aggressive treatments (though not all need invasive procedures) are hysterectomy, myomectomy, and embolization. While myomectomy and embolization look encouraging on the Web sites of the clinics that perform these procedures, each has risks[18] we were not willing to entertain. As to hysterectomy—well, not only did Tanya still want a child, my desire for grandchildren trumped any objectivity on my part about that choice.

Despite all the measures we had taken, Tanya's fibroid had caused a dangerous blood clot and was implicated in her anemia, blurry eyesight, general weakness, terrible pain, and other symptoms. It was stealing blood and energy from other parts of her body. It was also pushing against her liver, so her liver wasn't processing hormones correctly, creating emotional havoc, and she was looking yellowish much of the time. Not a pretty sight for a mother who was—wisely or not—still actively involved in her daughter's treatment.

I was starting to think of the fibroid as an independent force with a mind of its own. It wanted food, energy, blood, life force. It was out of control. The plant in the movie *Little Shop of Horrors* came to mind. It wanted to be fed. It was growing ever faster. There seemed no way to stop it. Its power was awesome. We reluctantly decided it was time to bring in the cavalry.

As fate would have it, Tanya was at one of my classes when we were coming to this decision, and one of my advanced students gave her a session. She told us about a doctor who she believed had saved her own life. This was no ordinary doctor. This was a woman who had gotten laws passed in three states regarding stricter requirements for informed consent prior to hysterectomies. This was a woman who had strongly advocated for more sensible practices in gynecology, invented less invasive surgeries for a number of female problems, and had been a thorn in the side of many of her more traditional colleagues. This was also a woman who had lost her license to practice medicine, perhaps for being a courageous pioneer, willing to utilize methods that were ahead of their time. But still, a revoked medical license is not the first qualification one usually insists upon when choosing a surgeon.

We met in a restaurant: Tanya, the doctor, and I. She quickly won our confidence. Her depth of knowledge permeated her answers to every question we asked. While I had learned very little I didn't already know from the other doctors and healers who had been helping with Tanya's fibroid, I learned a dazzling amount in a very short time about the issues and controversies in gynecology as she knowledgeably addressed each of our doubts and concerns. I was impressed by her humanity and caring for Tanya, for all her patients, for all women. She tuned in to Tanya's fibroid energetically, over the french fries, and she knew it as I knew it. With Tanya's permission, she laid her right hand over the area of the fibroid, and at one point turned to me so Tanya wouldn't hear and said, "She's really at risk!" She knew all our options, was able to quickly go through each of the alternatives our research had revealed, and discussed the reasons for and the liabilities of each. For instance, with embolization, what they tell you is that it is an ingenious procedure where silicone pellets are injected up through the femoral artery to cut off the blood supply to the uterus so the fibroid can't grow and ultimately shrinks because its blood supply has been cut. What they don't tell you is that cutting off the blood supply to the uterus can shrink and mummify it, so that within two years women often have to have a hysterectomy anyway. Beyond this, it is possible for the silicone pellets to go astray and permanently compromise the blood supply to other parts of the body.

After going through the available alternatives, each quite undesirable, the doctor described a surgery she had developed in which the fibroid is delicately separated from the uterine lining bit by bit and the uterus is then repaired. Because her license had been revoked, she told us about the hospital in Tijuana where she was currently doing her surgeries. A few days later, Tanya checked in. While part of me was concerned about a doctor who had been disbarred from practice performing an unknown procedure in a Tijuana hospital, I was feeling deeply blessed that I had stumbled upon this maverick of a physician who seemed to have examined conditions such as Tanya's from every conceivable angle and developed the most potent and compassionate ways of approaching them.

It was a tiny hospital with a capacity of only seven patients. Each private room had two beds, and a family member was expected to stay with the patient as advocate and support system. That honor went to me. Unlike American hospitals, the food was fabulous, cooked to order, and could be anything from gourmet dining to wheat grass juice. I prepared Tanya for the surgery by giving her extensive energy balancings several times per day in the days between the decision to proceed and the actual surgery. David had been giving her energy psychology sessions to prepare for the event emotionally. Besides addressing her natural fears about a highly invasive procedure and bolstering her confidence and positive expectations about the surgery, one of the themes of their work together was Tanya's lifelong reluctance to act decisively when someone, particularly a boyfriend, was hurting her. Cutting out a growth that was clearly doing great damage was a perfect metaphor for this issue. They focused on the part of her that did not want to interfere, that was racked with guilt about the thought of destroying this autonomous being growing inside her, no matter what harm it was causing. By the time of the surgery, she understood the metaphor perfectly and was feeling a warrior's determination to ward off anything attacking her or her body. Xena, television's beautiful warrior princess, had, in fact, been one of the figures she used in her mental imagery. I received from David my own energy psychology sessions as well, with one theme being that I not faint while watching my daughter being cut open.

On the morning of the surgery, besides balancing Tanya, I insisted on balancing the doctor as well. When the surgery began, I was there, scrubbed in with mask and gown, and there I remained through the whole four-and-a-half-hour procedure. The surgeon, a second doctor who was also the owner of the hospital, a nurse, and the anesthesiologist were also all in the room. The opening event—well, this was actually before the *opening* event—was conducted by the anesthesiologist. I was given the opportunity to energy test Tanya on the sedatives and local anesthesia that were going to be used. The choice and the dosages tested out well, though in other surgeries where I have been invited to participate, even when the choice of medication was correct, the amount was often too great. Not so here; the dosage was perfect for Tanya's body. I considered this a good sign about the medical team's intuition and attunement. To lend energetic support during the incision, I held my fingers about an inch and a half above Tanya's forehead, over the main neurovascular stress points, to keep her body from going into any kind of shock.

I thought for a moment that I wouldn't be able to watch the knife slice into my daughter's flesh, but I decided that I must force myself to witness the entire procedure. To my surprise, I wasn't a bit woozy. While I knew that the energy psychology had helped me prepare emotionally, a healing energy beyond anything I might have expected, and larger than any of us there, entered the room. We were suddenly in a loving, healing, sacred space. It was like an aura surrounding us, perhaps a product of the good doctors, the humble hospital, my love, and Tanya's spirit. It was palpable. I felt strong and glad to be there.

The cut was made. A large chunk of flesh had to be cut out in order to free the uterus. Directly beneath Tanya's skin, going four inches downward, was a thick layer of fat that had to be cut out. It was a creamy yellow and salmon color, with a healthy glow. I'd never thought of fat as I did in that moment. You saw instantly the necessity for fat in the body, to buffer your organs and bones, and why a woman is designed to carry more fat around her middle than a man. That fat protects the baby, as well as all the female organs. The shock to me was how beautiful the fat was, and how beautiful was the energy it emitted. It is a buffer that is there to comfort and protect. Society has given a bad name to this magnificent cushion. I've been much more accepting of my lust for Dairy Queens ever since.

The surgeon's hands disappeared into Tanya's body and reemerged with her uterus, now a hard, reddish-pink, glistening sphere slightly larger than a basketball. It was shiny and so taut it looked as if it were ready to pop. I instantly understood that there had been no time to waste—the need for this surgery had been urgent. The doctor lifted Tanya's uterus and placed it on her abdomen, attached now only by ligaments. Its energy was filled with a light that was so beautiful, it took my breath away.

When the knife cut into Tanya's uterus to begin to remove the fibroid, something shocking happened. With the first slit into the uterus, an energy sprayed out that was exactly the opposite of what I had been seeing. Rather than a beautiful healthy glow, it looked like an evil dark force. I didn't have any visual cues from looking at her uncut uterus, even its energies, that such a force was held inside it. Not only was the energy ugly, so was the fibroid that was to be removed. It was one of the ugliest things I'd ever seen. I immediately had the thought that it held years of psychological trauma and pain. It was a terrible energy; it looked absolutely malevolent, both as a red, disorganized blob and in the energies it was emitting.

Tanya's energy, by contrast, looked so pure. She was an innocent lying there, while a negativity that seemed absolutely evil had lodged and taken a strong hold in her body. It came to me in a flood that the surgery was the most holistic thing we could do, to cut out this darkness from Tanya's body, from her life. The surgeon, on first seeing the fibroid, said, in memorable understatement, "I'll bet Tanya's been in a bad mood for a long time."

To remove the fibroid in one piece would have required a cut that the uterus could not have survived, so the fibroid was cut in two. The first part to be removed was as big as a huge melon. It adhered to the inside of the uterus, so the surgeons had to very delicately cut it away. The care that went into this—small snippet by small snippet—was astounding. While Tanya's uterus was sitting on her stomach, the doctors had their instruments inside it, like working in a tiny cave. Their painstaking steps were ensuring that Tanya's uterus would be preserved. The whole process was amazingly respectful of the integrity of her body.

When the last of the fibroid had finally been cut away from the uterus, along with some 25 fibroid "seeds," everyone in the room took a big sigh. Hours of care and precision had succeeded in separating and removing the invader. Throughout the surgery, clamps had been used, so there was virtually no bleeding. I was awed by the skill and care I was witnessing. Except for brief, soft communications, the operating room had been in a holy silence throughout this battle to excise the fibroid without destroying Tanya's uterus. Now lively, light classical music was brought in as the process began of stitching the uterus back together. It was not just a matter of closing the incision. Wounds remained wherever the fibroid had adhered to the inside lining of the uterus. So the doctors were now sewing the inside of the uterus, determined to leave it strong enough to carry my grandchildren. At various points along the way, the surgeon explained to me why she was choosing a particular procedure rather than an alternative for the purpose of preserving Tanya's fertility.

The two doctors stitched and stitched, reminding me of a seamstress and a tailor. My appreciation was immense. I myself am a seamstress. Sewing is my relaxation and my meditation, and here I was watching my daughter's womb being ever so skillfully and patiently sewed, like a beautiful multilayered quilt being hand-stitched. After an hour of rapt attention and awe, I said, "You must have put 300 stitches in there." They laughed. "Closer to 500!"

When the uterus was completely sewn up, the lead surgeon stood back, looked at the other doctor, nodded her head, and smiled. This was his cue to lift the uterus from Tanya's stomach and cup it in his hands. His arms were outstretched and he placed his head in his forearms. All was silent. The music had been cut off. No one spoke. He continued in this posture for what I would guess was at least three minutes. Then he stood up, took a breath, and placed the uterus back on Tanya's stomach.

At this point, the surgeon told me, "He is helping the cells in her womb remember what it was like before they were harmed." She explained that "cells hold memories!" (as if this weren't one of the core premises on which my career is built). Then she nodded to him, and he again lifted and held the uterus for what seemed to be another three minutes. When he did this, everyone else in the room was in a quiet reverence. I took photos of the entire operation, and one of the most beautiful is of the doctor holding Tanya's uterus. It had a beautiful glow within and around it that everyone could see and that is evidenced in the photo.

This was a most sacred experience, like being at a holy altar during a high spiritual ceremony. I told the doctor that it looked as if he was praying. Taken aback, he said, "Well . . . I was." Then they placed Tanya's uterus back in her body and did the final stitching. When the surgery was completed, Tanya opened her eyes and smiled at me. As we had planned, I began to give her another energy session to help her body restore after the trauma of surgery. One of the givens after a surgery is that the patient's energies will be running in a homolateral pattern. This occurs, actually, any time a person's health is compromised, a way that the body conserves its energies. But it is harder to function and harder to heal while the energies are in that pattern. So one of the first things to do after a surgery is to begin to redirect the energies from a homolateral to a crossover pattern. I've never seen anyone right after a surgery who was not homolateral. To my amazement, Tanya was not! Her body was so prepared for the surgery that her energies just kept on flowing! I was elated! During Tanya's recovery, a variety of medications were prescribed for pain and swelling, and when some of them energy tested weak, the doctors were good enough to change the prescriptions until medications and herbs that were compatible with Tanya's energies were found.

Had we not been led to this unusual surgeon, Tanya almost certainly would have been forced to have a hysterectomy, and in any case, would have lost the possibility of carrying a baby. Now, at the time of this writing, it is still a choice she can make. But whether or not she ultimately carries a child, I am thankful her uterus was preserved. The full consequences of a hysterectomy are often not disclosed to women. For instance, the female organs are tightly arranged so that removing the uterus, even if it is prolapsed, may itself cause a prolapse, a falling out of place, of another organ. One of my friends wound up in a wheelchair, unable to walk, due to complications from a hysterectomy. While the procedure should never be done casually, hysterectomies are performed almost routinely in the United States. They are carried out at nearly double the rate of that in England or Europe,[19] suggesting that social and economic forces within the medical system are a strong factor in determining whether a hysterectomy is to be performed.

Nonetheless, my indebtedness and gratitude for the tools of modern medicine were immense around the time of Tanya's surgery, and they will never be forgotten. Even though my professional role seems to have become one that challenges the medical profession to expand its paradigm and soften its arrogance and rigidity, when surgery or pharmaceuticals are required and skillfully administered, I can only watch with awed respect that humanity has devised such marvels of compassion and healing. Tanya was, of course, very fortunate to have found a physician who also thinks outside the box, but ultimately it was her surgical skills that saved the day. I, in fact, have always viewed energy medicine and conventional medicine as complements to each other rather than competing approaches. Many of my cries of protest have been pleas for conventional medicine to come to that view as well. For instance, if you prepare your body energetically prior to surgery, and attend to it energetically following surgery, as I did with Tanya, the chances that your body will respond to the surgery as intended are enhanced, and your chances for a robust and speedy recovery greatly increased. When I revise *Energy Medicine* for its tenth-anniversary edition (how time flies!) as soon as I complete this book, I plan to add a section on how to apply energy methods to support necessary but invasive medical procedures such as surgery. I'm thinking of calling it: "If surgery becomes necessary: The cutting edge as a last resort."

Pregnancy

Pregnancy is the miracle that occurs after two microdots of protoplasm, each the product of an independent universe, implausibly find their way to each other and merge into a new microdot. This new microbeing carries all the necessary information for creating a trillion-celled organism that hosts the most sophisticated computing system known, capable of building skyscrapers, writing poetry, exuding love, and transcending itself in mystical rapture. Much of this information had been carried in the energy fields of the egg and the sperm, and it is now in the *energy field* of the new microbeing (called a zygote), rather than in its *genes,* but that is another story, laid out in Chapter 1. Discussed here is the growth of the microbeing into a human baby, a play that unfolds in the theater of your womb. The stage management you provide, both physically and energetically, will impact the production in ways you already know and in some ways you might not. Following is a guide to the energetic aspects of that stage management.

Pregnant women came into my office for a myriad of reasons. Some came with concerns about the health of their baby. Some came because they were fearful of a Cesarean section and wanted to work with someone who would support a natural birth. Some came because they were swelling or suffering with morning sickness or other symptoms. Some wanted me to tell them the color of their baby's aura. But all desired to stay very healthy during their pregnancy so their baby would have an auspicious entry into the world.

Supporting the Health of the Baby. Whatever the mother does to energetically stay in balance helps the baby. Do the basics laid out in Chapter 2, particularly the Daily Energy Routine and all the techniques that stretch your body and that keep your energies crossing over. As with everything else, but particularly with pregnancy, keeping spleen meridian strong is vital to your health. When a woman becomes pregnant, one of the earliest energetic changes to occur is the intensification of her spleen energies. Each meridian has a pulse that can be felt on the wrist, and one of the ways I know that a woman is pregnant is by checking her spleen pulse for otherwise unexplained increases in her spleen meridian energy. This increase is not surprising. Spleen meridian takes on a host of new tasks oriented toward nurturing the baby. So you want to be sure it has lots of reserve energy. The spleen/triple warmer balancing sequence suggested to promote fertility (page 180) can do much to keep spleen meridian strong and balanced. The other meridians that are most intimately involved with fertility—the kidney, liver, stomach, and circulation-sex meridians—also have important roles in fostering a healthy pregnancy. Methods for supporting them are suggested below. One of my favorite ways to keep mother and baby connected follows.

Heart-to-Womb Connection
(time—about 9 months)

One of my favorite ways to support the mother–baby bond is to bolster the energetic connection between heart and womb. Standing or lying down, take a few deep breaths in through your nose and out through your mouth. Place one hand over the middle of your chest and the other over your womb (Figure 5-7). After a few more deep breaths in through your nose and out through your mouth, breathe naturally. The breathing instructions help ensure a healthy flow of energy, blood, and oxygen between the lungs and the womb. And the power of your love toward your baby is laser-focused.

Paving the Way for a Natural Birth. A pregnant woman whose first two children were delivered by Cesarean section came to me with the hopes of being able to deliver her third child naturally. Her obstetrician was amused by the techniques I gave her. He told her these were good to do to keep her relaxed, but that she should not fool herself into thinking she would not have to have another Cesarean section after already having had two. He insisted, in fact, that she could not continue to be his patient if she did not abandon her plan for a home delivery. A home delivery in her case felt foolhardy and dangerous to him. I, however, was encouraged by her progress, by the way the energies were no longer blocked in the areas where she had scar tissue from the first two Cesareans, and by how her whole abdominal area looked healthy and strong. I told her that I felt confident she could deliver naturally. She, understandably, bowed to her doctor's recommendation and dropped the plans for a home birth. When the time arrived, however, there was no choice to be made. The baby came about three days before the doctor expected, and with almost no warning. Labor came on suddenly. She was in bed. Her husband was home. They called the doctor and began preparing to go to the hospital. She made it as far as the liv-

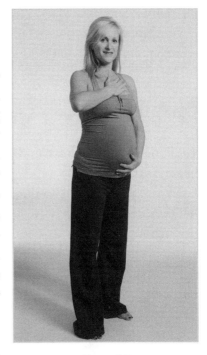

Figure 5-7
HEART-TO-WOMB
CONNECTION

ing room. She squatted, and the baby came out. It was a remarkably easy birth. No complications. The doctor, to his credit, put substantial effort into finding out what had happened and wound up learning a great deal about energy medicine.

Healthy Pregnancy Sequence (time—less than a minute)

To support a natural birth and decrease the chances of needing a Cesarean section, begin with everything recommended in the discussion under Supporting the Health of the Baby. In addition, to support the energies most involved with pregnancy, do the Healthy Pregnancy Sequence on a daily basis. Breathe deeply throughout the entire sequence.

1. Push your thumbs into the points at the top of your cheekbones, directly beneath your eyes (Stomach 1 and 2). At the same time, lightly place the pads of your fingers on your forehead, on the bony area about two inches above your eyebrows (see Figure 5-8). Hold this position for at least a minute while breathing deeply. This part of the sequence can be done independently of the other two. Do-ing it several times per day reduces stress, relaxes the abdomen and uterus, and gives energy to the legs to support your growing body.

2. Place your fingertips on the arch of your pubic bone and massage deeply the areas under each finger (these are neurolymphatic reflex points that open the energies in your entire pelvic region).

3. Start with the middle finger of your left hand on the center of the pubic bone; move it left about two inches and drop it about one inch, staying in the groin crease. Press and gently massage the point there

Figure 5-8
HEALTHY PREGNANCY
SEQUENCE

on your groin (Liver 12). With your right hand, massage the K-27 points (page 53). Repeat on the other side, massaging the liver point on the right side with your right hand and K-27 with your left hand.

Calming the Circulation-Sex Meridian. During the last trimester of pregnancy, calm the circulation-sex meridian by having a friend or partner hold its sedating points (page 148). This helps your circulation, supports your own heart and your baby's heart, and reduces the lower back pain and exhaustion that can result from carrying the baby. If no one is available, and if it is difficult for you to reach the sedating points, gently hold the "devil's horn points" (about three inches above the ears) for two to three minutes. These are circulation-sex neurovascular points.

Balancing Your Root and Womb Chakras. All of these procedures will help your energies flow and your hormones stay balanced so that your body's natural, organic rhythms take over in preparing you for the birth. Another extremely valuable procedure is to clear your chakras, particularly your root and womb chakras, the spiraling energies directly over your pubic bone and over your womb. An entire chapter of *Energy Medicine* (Chapter 5) is devoted to understanding and working with your chakras, but here is a brief version of the procedure. While you can do this on yourself, it is much nicer to relax into it as another person does it for you. The Five-Minute Daily Energy Routine (page 51), or at least the Neurolymphatics, Crown Pull, and Hook-up, are good preliminaries. Then, circle either hand in a counterclockwise direction two to three inches above your pubic bone. Circle slowly for at least a couple of minutes, breathing deeply. Shake your hand off and circle over the same area in a clockwise direction for a minute or two. Repeat the same counterclockwise and clockwise circling over the area of your womb, just below your navel. End with figure-eight patterns over the entire area.

Goddess Belly Hold. This technique keeps the energetic connection between you and your baby strong and clear by keeping your belt circuit in a good flow. The belt circuit, or belt flow, is a radiant circuit that surrounds your waist and, when you are pregnant, also moves below your belly, supporting the growth of the baby. It connects the energies of the top and bottom parts of your body, and allows a

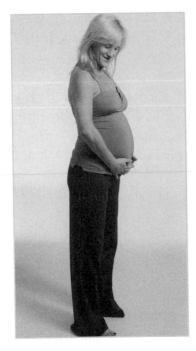

Figure 5-9
GODDESS BELLY HOLD

strong flow of blood, lymph, and oxygen to move through the womb and assist a healthy birth and baby. This vertical distribution of the energies is critical to the belt flow when you are pregnant. Place your hands at the sides of your waist with your thumbs on your back. Slowly and firmly pull your hands forward and bring them downward until they are cupping the bottom of your womb (see Figure 5-9). Take a deep breath and then slowly let go. Repeat at least one more time. This procedure feels wonderful for most people, and it reestablishes a top-to-bottom harmony in your body. If you can get another person to do this for you, it is a joy. You can lie down on a bed so the other person can place each hand at the center of your back and pull you up from behind your waist. As the person's hands pull out to the sides of your body, you can relax again into the bed.

Breech Position. A Cesarean section is often performed when the fetus is in a breech (feetfirst) or transverse (sideways) position. But energy work can sometimes get the baby back into the headfirst position that nature intended. The energies of the unborn child are distinct from those of the mother, and when you do a chakra balancing on the mother, you can feel the mother and the child responding independently. Feeling the child's response always makes me happy. Sometimes a baby is in no rush to get out, and if you relax the womb area with a chakra clearing, the baby will actually begin to follow your hand. In one case, a woman was overdue for delivery, and her child was in a breech position. The mother was paralyzed with fear, and did not want to have a Cesarean section. She asked me if I could find anything in her energy that was keeping her expected daughter, already named Melissa, from turning to the right position. Melissa's energy appeared healthy, but she was in no rush to get out. The only problem I could find was that the mother's abdominal muscles were tightly constricted from stress. I relaxed this area with a chakra clearing over her womb. After about two minutes of counterclockwise motion, it looked and felt to me as though Melissa's energy was getting in sync with mine and picking up the rhythm of my hand. After about fifteen minutes, she began to turn, as if following the energy of my hand, until she was upside down, ready for birth. Another person in the room was jumping up and down and screaming, "I can see her turning!" The mother went directly from my office to the hospital, where she gave birth to a beautiful baby girl. While you cannot always count on this technique to prevent a breech delivery, I have had nearly identical experiences several times. Also, since first describing this method in the original edition of *Energy Medicine,* a number of midwives and nurses have told me that the technique led to the repositioning of a fetus, making a natural delivery possible.

Miscarriages. Again, everything mentioned to this point supports a healthy birth: the Daily Energy Routine; enhancing crossover patterns; stimulating and balancing the spleen, triple warmer, stomach, kidney, liver, and circulation-sex meridians; balancing your root and womb chakras; and keeping your belt circuit flowing. Also of particular value in preventing things from going awry are techniques that keep the energies connected, such as the Hook-up (page 62) and Connecting Heaven and Earth (page 44). It is also important to accept that sometimes a miscarriage is nature's way of releasing a baby who is damaged or otherwise not right to live in this world and that nothing you could have done was going to change that.

Three common problems during pregnancy are morning sickness, back pain, and toxemia or preeclampsia. Each can be addressed by energy medicine, often in partnership with other approaches.

Morning Sickness. Bouts of nausea and vomiting affect between 50 and 95 percent of women during the first trimester of pregnancy. While not harmful to mother or child (unless they lead to dehydration or malnutrition), they are no fun either. Folk remedies that are helpful to some women include eating six smaller meals instead of three large ones, getting plenty of rest and good exercise, drinking fluids 30 minutes before or after meals but not with meals, sniffing lemons or ginger, and taking about 200 mg of vitamin B_6 daily. A variety of energy techniques can also help bring morning freedom.

1. The quickest and most direct step you can take for morning sickness is the Hook-up (page 62), placing your middle fingers at your navel and third eye, pushing in, and pulling up. Hold for at least three deep breaths. This hooks up the central and governing meridians so that energies can realign.

2. A second intervention is to squeeze the bottom corners of one big toenail while your other hand pushes up under your cheekbone beneath your eye. Sit this way for a minute or two, breathing consciously. Repeat on the other side. These two techniques will alleviate morning sickness for many women. If they do not work for you, the following helps with the most common patterns. None of them will hurt you, so experiment and see which works best for you.

3. Rub your hands together until you feel warmth. Place one hand where you feel discomfort and the other hand on the back of your neck. Breathe deeply and hold for a couple of minutes.

4. Become as comfortable as you can in a chair. With one hand, grasp with your thumb and middle finger on the back of your foot, just behind your ankle. Squeeze the area behind your ankle on either side of the Achilles tendon. At the same time, with your other hand, place your middle finger beneath your cheekbone and push upward (see Figure 5-10). Hold for a couple of minutes. Repeat on the other ankle and cheekbone.

5. Hold the first set of small intestine meridian sedating points (page 145) for about two minutes, and then lay the heels of your hands against your cheekbones, with your thumbs on your temples and your fingers on your forehead, for about a minute.

Figure 5-10
Good Mornings Position

At least one or two of these should work quite well, and that is all you need to use. Experiment to find out what works for you. The feedback will be almost instantaneous.

Back Pain. The weight of the growing baby places tremendous physical pressure on the expectant mother's back.

1. Having your partner or a friend do a Spinal Flush daily (page 60) prevents stagnant energies from accumulating as the back does double duty, and this keeps energies moving through and strengthening your back.
2. If you are feeling pain around your kidneys, sedating the kidney meridian (page 145, following the special instruction there for pregnant women) releases tension in the kidney itself. If you are more than a few months pregnant, you probably will not be able to reach the points on your feet and will need to ask your partner or a friend to hold the points. If sedating the kidney meridian did not alleviate the pain, sedate the large intestine meridian (page 146), which governs the muscles at the waist in the back.
3. If the pain is in your gluteal muscles or sacrum, sedate the circulation-sex meridian (page 148). Again, you will need your partner or a friend to hold the points.

Toxemia. Toxemia, also called preeclampsia, is characterized by high blood pressure, swelling that will not subside, and large amounts of protein in the urine. All the techniques suggested above, such as maintaining a balance between the spleen and triple warmer meridians and keeping the liver and kidney meridians in a good flow, help prevent or reduce toxemia. The Five-Minute Daily Energy Routine is usually enough to maintain these balances, though sedating triple warmer, liver, and kidney meridians, and strengthening spleen, gives an extra boost. Massaging the lymphatics (page 59) can also further reduce toxemia.

Delivery

My dear friend Sandy Wand, one of Ashland's legendary midwives, tells me that at about eight months into pregnancy, it occurs to the expectant mother that the "microdot" (she had just read a draft of the pregnancy section above), which has now become a beach ball, is going to have to find a way out through an opening that suited the microdot but seems quite inadequate for the beach ball. Where all the awareness had been focused on the pregnancy and growing baby, now there is a shift of awareness that this pregnancy is about to lead to a birth. During the last weeks of pregnancy, a natural energy comes in to prepare for delivery, changing physiology and consciousness. The pelvic floor is spreading. The baby is dropping into it. The baby's head becomes engaged in the pelvic floor, and the baby settles into the position in which it will be birthed. There is no escape, and biology by-passes all theories about how this is going to happen.

Fortunately, the body *knows* how to give birth. And it can be supported in many ways. Simple techniques such as the Hook-up (page 62) or having your partner or friend press in for about a minute at the power point (the center indentation where the head joins the back of the neck) help calm the nervous system and reorient the body's energies.

During labor, contractions create tension. Energy can be pulled off the body between contractions by simply smoothing with flat hands down the body and off the feet. Because labor is such an intense physical and emotional experience, gently placing your hands on the neurovascular points on the forehead or behind the knees (some women do not want to be touched on the head during labor) can relax the body, release anxiety, and calm the mind so that the body can do what it knows how to do naturally. This is a simple intervention that may yield major benefits. Reducing tension and panic allows the body's organic wisdom to operate unimpeded. Gently smoothing the forehead with a damp cloth stimulates the same neurovasculars.

Placing your hands on the woman's sacrum with pressure during contractions helps keep a strong current moving in the radiant circuit called the penetrating flow. This taps into and supports the womb's powerful energies, aligns the mother with the energies of the baby, reduces pain in the lower back, and harnesses all the radiant circuits, the natural resource energies for delivery. Holding the sides of the woman's feet early during labor connects all the body's energy systems and helps them coordinate with one another. A method called the Brazilian Toe Technique can be enormously comforting during labor, bringing greater efficiency to all the body's energy systems. When you do this for someone, the recipient is lying on her back. Be sure to also find a comfortable position for yourself.

Brazilian Toe Technique (time—10 to 15 minutes)

1. Position yourself at the feet of the mother-to-be. Place your thumbs beneath her middle toes and your middle fingers on top of the same toes, on the toenails (see Figure 5-11). Hold lightly for two to three minutes. Breathe in through your nose and out through your mouth.

2. Slide your thumbs to beneath her fourth toes, then slide your fourth fingers on top of the fourth toes. Hold lightly for two to three minutes. Continue the same breathing.

3. Slide your thumbs beneath her little toes, then slide your little fingers on top of the toenails of her little toes. Hold lightly for two to three minutes.

4. Slide your thumbs beneath her second toes, then slide your index fingers on top of the toenails of her second toes. Hold lightly for two to three minutes.

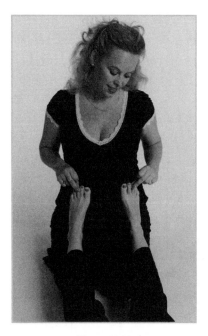

Figure 5-11
Brazilian Toe Technique

5. Slide your thumb beneath her big toes, then place your index fingers and your middle fingers so that one is on each side of the base of the toenails of her big toes. Hold lightly for two to three minutes.

Every physical and mental resource a woman has is pushed to and beyond Olympian proportions during labor. In giving us a skull large enough to contain a human brain while streamlining our pelvis to allow us to walk on two legs, nature was pressed to make some ingenious compromises as well as some sacrifices in evolving the modern human, and easy childbirth was one of those sacrifices. A culture's practices in counteracting this big-brain/small-pelvis dilemma tells a great deal about its compassion, wisdom, and attunement to the feminine principle, a topic I will skip before I go into a diatribe of anguished bewilderment. Suffice it to say that the midwife tradition has much to teach us, and a few Western hospitals are making great positive strides. And whatever the birthing environment, wherever nature has made significant biological compromises, things can go wrong.

It is beyond the scope of this book to go into the emergency procedures that may be required during a birth, and Western medicine excels at saving lives in desperate situations. Yet even here, less invasive energy techniques are available as a first response, and they can often prevent the need for more Draconian measures. For instance, if there is weakness or too much variability in the fetal heartbeat, holding the acupressure strengthening points on the mother's heart and spleen meridians, or tracing her heart meridian, will often resonate between mother and child, strengthening and stabilizing the baby's heartbeat and decreasing the likelihood that medication will be needed. If, after a certain level of dilation and contraction activity, the labor stalls, techniques for getting the energies grounded and moving—such as squeezing and massaging the sides and tops of the feet, holding and swaying the mother in figure-eight movements, or simply getting her to move her body—will often spontaneously restore progress.

When I went into labor with my daughter Dondi, my doctor, Paul Brenner, M.D. (a legend in our family about what a doctor ought to be), was with another patient in a different location and couldn't get to the hospital for a couple more hours. He was able to contact one of his friends, and he got him to agree to immediately go to the hospital to be with me. He gave him two directives, to hold me when contractions came and to keep the nurses away from me until Paul arrived. This man not only wasn't a doctor, he showed up in a football uniform, cleats and all. Whenever a contraction would come, he would hold me as instructed, and he also instinctively rocked me in a figure-eight manner. It was incredibly comforting. My body would relax in these big arms. I felt totally taken care of. When Paul arrived, he took over these duties, also instinctively rocking me in a figure-eight motion. He asked me what would make this the most wonderful birth experience possible. I said that I would really like to be able to watch the birth. A couple of hours later, a giant mirror was installed on the wall of the room where I would be giving birth. With a beaming smile, Paul said, "Take a look, Donna! Happy 'birth' day!" After the delivery of a beautiful baby girl, Paul was crying openly and hugging me and my perplexed husband. A few years later, after no contact, Paul called me at home about an unrelated matter and three-year-old Dondi answered. He said, "This is Paul Brenner. Is your mommy there?" Dondi squealed, "Paul Brenner! I'm your baby!"

. . .

Female sexuality, fertility, pregnancy, and delivery are a woman's distinctive birthrights. The culture in which a woman lives maps a route through each of these profound journeys. The maps provided by our culture have become increasingly estranged from our most fundamental energies, natural rhythms, and biological foundations. (I hope every woman who is considering becoming pregnant or is early in her pregnancy will watch the documentary film *The Business of Being Born,* with Ricki Lake, and take steps based on its counsel.) Outside our cultural frameworks, however, are maps that are more life-affirming, more keyed to our highest potentials. As we fully grasp the situation, we are pressed to become cartographers, mapping old wisdom into new territory that holds both opportunities and threats our grandmothers could not have imagined. We are, through the very lives we are living in these challenging times, changing the maps and fueling hope for the next generation about what is possible for a woman. I hope the tools presented in this chapter support you in these grand adventures.

Chapter 6

Menopause—Gateway to Your *Second* Prime of Life

Menopause is not just the end of fertility and childbearing years
that are so often seen as women's prime. It is the time for a
woman to take charge of her health and make the changes
that will carry her through her second prime of life.

—BERNADINE HEALY, M.D.
Former Director, National Institutes of Health

Menopause is a journey of mythic proportions.[1] It is the gateway into your second biological prime! My intention with this chapter is to help you navigate through that gateway into a *richer* second prime. Let me know if I succeeded: www.ed-em.com/2ndprime-survey.htm.

Until you get there, it is hard to imagine how strange this territory can be. Since I knew PMS at its worst, I thought I would sail through menopause. You sometimes get off easy on one if the other is dreadful, much as I had enjoyed easy pregnancies after monthly PMS agony. Women like me, who are rich in estrogen and thus have difficult periods, often have an easier time of pregnancy and menopause because the ratio of progesterone, which we are low on, increases in relationship to estrogen. I, however, turned out to be a victim of my own expertise. Natural progesterone had so firmly become my drug of choice by serving me magnificently during the PMS times that I kept taking it into menopause, and I threw off my progesterone/estrogen balance. So I wound up with the biochemistry of a more typical menopause instead of taking advantage of my natural excess estrogen. I had the experience for the first time of drenched sheets, insomnia, and a feeling of deadness, none of which would have occurred if I had not overdosed on progesterone. Sometimes you just can't win. But you can learn from my mistakes. Read on.

Dozens of books on menopause can be found in any major bookstore, but to my knowledge this is the first whose approach is based primarily on energy medicine. Several, however, are excellent compilations of conventional medical as well as folk remedies for common symptoms (e.g., Wingert and Kantrowitz's *Is It Hot in Here?*[2]). I have also loved five masterpieces that address the topic from their own unique angles, written by Leslie Kenton, Ph.D.,[3] Susun Weed,[4] Christiane Northrup, M.D.,[5] John Lee, M.D.,[6] and Susan Lark, M.D.,[7] respectively. While it is beyond the scope of this chapter to attempt to cover all the material in these superb resources, some of the essential ideas that provide a context for an energy-based approach to menopause follow.

Coming into your second prime requires a major rearrangement of the perspectives and values that served you earlier in your life. Many of the books about menopause focus on the psychological challenges, which are substantial. Our focus, however, is on energy tools that can help you move through menopause with greater ease and in better health. This is the best concrete preparation I know for the psychological and spiritual as well as the physical journey of menopause. Energy medicine helps you at all levels.

PREPARING BODY, MIND, AND SPIRIT FOR THE JOURNEY

Every emotional challenge is a physiological experience. Trying to think your way out of depression or anxiety usually does not work. Our culture may tell us that will and determination are all that are needed, but it would do much better teaching us how to move our energies when they are stuck rather than teaching us to blame ourselves for *being* stuck. It is not "all in your mind"; it is all in your energies. The emotions of menopause reside in the chemistries of our bodies, and they are powerful. At the same time, massive, cataclysmic shifts are retooling your body from a baby-making machine to an independence you never before could have imagined. While nothing is going to stop you from going through this tumultuous process, energy medicine can deeply support you during the transition and help you come out of it healthier, happier, and more independent. By becoming adept at balancing your energies, you are better able to navigate your way into the gifts of menopause: empowerment, self-reliance, and freedom from the compulsion to cater to others' needs or agendas for you.

My own passage through menopause was great training for being attuned to the women who consulted me in my practice. After you have served as your own laboratory, you get much better at guiding others. As it happened, most of my closest friends started through menopause about four years after me. Still in the experience myself, but a little further along, I was able to resonate deeply with their complaints, terror, and confusion.

One day I dropped in on a friend who was known for being extremely fiery and opinionated, a powerhouse personality. I hadn't heard from her in a while and wanted to check in. She casually told me, in a dry humdrum voice, that she no longer cared about me. It wasn't personal at all, she assured me. She didn't care any longer about anybody or anything. The well of her human passion, along with her compassion, had dried up. She said she didn't desire a thing. Her hearty sexual appetite was gone. She felt dead to all of life. And she hated her itchy, flaky, dry skin. When I offered to give her a session to help her deal with her dead emotions, she was eye-roll irritated and felt patronized. She was, however, willing to let me help her with the torture of her itchy skin.

So I had her lie down on the floor, and we got started. I energy tested every meridian in her body that might be involved with skin conditions and corrected each imbalance. By the end, her skin wasn't itching anymore. This got her attention. The relief was so great, in fact, that she wanted more. We went to work on her hormones. While several energy systems were involved—such as her chakras and radiant circuits—her liver meridian turned out to be the key. Deep energies that had seemed lost to her were again available. At one point, to her surprise, she began to sob. She had believed she would never feel enough to cry again, and her tears seemed to open a floodgate for all the liquids in her body. She could cry again. She could feel sexy again. Her dead mood began to lift, and all of life began to seem juicy. Over a number of sessions, these shifts became more stable. She no longer felt "dried up." While her body would continue for some time to go in and out of chemical imbalances, the fluctuations became far less extreme, and with new energy tools in her back pocket, menopause became a bothersome challenge rather than a hopeless descent into the void.

It is striking to me how the decade between the years of approximately 48 and 58 often determines whether a woman finds her way into a vibrant second prime—strong, healthy, and with higher self-esteem than before—or a period of rapid aging, as if the woman's life is winding down. While many psychological and physiological factors are involved, the common denominator seems to be that the women who find their way into a second prime actively engage their body's energies (whether or not they are specifically thinking in those terms), not accepting at all that this is a downward spiral. Those who have come to me for help presented symptoms associated with menopause, but their journeys included hearing what their bodies were really trying to tell them—in their native language, energy—and aligning themselves with their bodies' requests.

Menopause as Nature's Plan for You

From maiden to mother to crone: the archetypal phases of a woman's life. The spirit of the crone arises in the postmenopausal woman, according to Leslie Kenton, and becomes "deeply and spontaneously sexual, assertive, straight, incorruptible, prophetic, intuitive, and free . . . the qualities that most terrify the patriarchal culture in which we live."[8]

In the midst of my own menopause, with its multiple emotional and physical challenges, a new energy slowly began to emerge, a force that seemed to come up from the earth and permeate my cells. It was a power I'd not known before. I was new to this realm. I was in undeniably unfamiliar territory with novel demands and newfound empowerment. While I wasn't exactly thinking in these terms, I was entering the world of the crone. I liked it. Despite the emerging challenges, I felt more whole, and certainly more assertive. I was free in ways I'd not known before. A silly incident stands out in my memory as an emblem of this new assertiveness. I was crossing the street, lost in my thoughts, and an impatient young man, behind the wheel of a shiny new convertible, "flipped me off." I guess I was taking too long to get across. While in the past I would have felt terrible for having inconvenienced him, I was now totally amused. How stupid it seemed for him to spend his energy and upset on something so trivial. I turned to him and said with calm assertiveness, "Hey, I'm fifty years old. I don't have time to go into that space. And you don't, either!" As his eyes met mine, his hostile expression turned into a quizzical look, and then it was as if he really saw me for the first time. He said, "No, we don't, do we? Thank you!" By fully embracing the way of her own heart, the crone invokes the power to change the way others think about their choices.

After half a lifetime of subduing your more independent qualities to serve husband, children, and society, you are ready to be subdued no more. Your goals become clearer, your mind more focused, and your power unleashed. The wisdom of innumerable experiences congeals, and you find yourself to be a teacher and leader in your extended family and in your community. That is nature's plan for you. Are you ready?

This plan, for a second prime, is supported by the hormonal changes that correspond with menopause. It is programmed into our genes. Our sexuality is another remarkable evolutionary achievement. No other creature on Earth, male or female, comes close to having our experience as sexual beings.

According to Leonard Shlain, "a yawning chasm separates the reproductive life [of human females] from that of the females of the other three million sexually reproducing species."[9] We are the only female species capable of sexual union year-round rather than only when we are "in heat." We are the only female species with such dramatic menses or with periods that are synchronized with the moon or that discharge so much blood that we are, in the wild, all but dependent on animal meat to replace the iron we lose. We are the only species with a brain that is capable of overriding the circuitry that demands obedience to our sexual urges, allowing us to refrain from sexuality for the purpose of choosing when and if we become pregnant, and also allowing us to be highly selective in our choice of a mate. In contrast, a female chimp mates, on average, 138 times with thirteen different males for each infant she births.[10] We are the only female species capable of prolonged and multiple orgasms of such intense and sustained pleasure. We are the only female species whose architecture contributes to extreme pain during childbirth. And we are the only females to experience menopause so early in relationship to our potential life span.

Shlain's rather extraordinary hypothesis is that these distinctive qualities of human female sexuality combined to drive human evolution forward so we would thrive by intellect rather than brawn. We can't outrun or outclaw a lioness, but we can outplan her! Foresight—"the ability to maneuver conceptually in the dimension of time"—ultimately provided us with control over no less than the earth's destiny. Women's sexuality, according to Shlain, "taught our species how to tell time."[11] Connecting the lunar and menstrual cycles, noticing the relationship between intercourse and the cessation of menses, and observing that babies appear nine moons after the last menses were all training grounds for beginning to understand the workings of time. Many of these reproductive oddities were actually compromises. They did not, in themselves, advance our biological fitness. But they combined to show us how present events relate to the past and to the future. Mortal risk at childbirth, for instance, increased exponentially because of the biological compromises that were necessary when our skull size increased to accommodate a large brain and our pelvis was redesigned to allow us to walk on two legs. There was, in fact, a period of history when so many women were dying in childbirth that our entire species was at risk. What was the evolutionary advantage in that? But we survived because women figured out the relationship between sex and subsequent pregnancy, and began to take control of their reproductive lives.

Discussing menopause, Shlain reaches back to Hecate, the prototypical crone of Greek mythology, recognized by both men and women for her freedom, courage, strength, wisdom, and magical powers. Although the crone certainly earned the reverence she received by adeptly distilling a lifetime of experience, another major factor contributes to her potency. This, according to Shlain, involves "the dramatic realignment of the serum concentrations of her estrogen, progesterone, and testosterone."[12] While menopause launches a steep drop in all three hormones, the proportion of testosterone in relation to the other two jumps sharply. Estrogen production often decreases 70 to 80 percent following menopause; progesterone production decreases by more than 99 percent; but testosterone production decreases by only 50 percent or less. This means that the amount of testosterone relative to estrogen often doubles (in some cases, it increases twentyfold) and, relative to progesterone, it increases at least fiftyfold. With these radical changes in the balances among the hormones that control personality come dramatic shifts in a woman's psyche. The indecision and pliability that often mark a woman's youth "are replaced by clearheaded assertiveness."[13] And when mature women, "brimming with testosterone (relatively speaking)," are freed from child-rearing duties and reenter the world on their own terms, they "begin to exert a wider influence on the welfare of the society."[14]

As Margaret Mead once said, "There is no greater power in the world then the zest of a postmenopausal woman." During our first prime, we are wired to bear children and nurture a family. Then, in our second prime, we are wired to lead our communities with passion and zest—using all the pragmatic ways of loving and caring we learned during our first prime as training for the job. This is, in my humble opinion, much better training for community leadership than is a political science degree.

Menopause: New Chemistry/New Energy

The transformation from a biochemistry that is oriented toward hosting the monthly drama of releasing an egg and preparing it for conception to a biochemistry supporting newfound independence is indeed jarring. In addition, estrogen and progesterone serve many functions beyond promoting sexual desire and supporting pregnancy. The effect of these hormones on secondary characteristics—from skin texture to hair luster to bone strength—also goes into turmoil at menopause. Seventy-five percent of the 40 million menopausal women in the United States complain of symptoms associated with these changes.

The appeal of hormone replacement therapy (HRT) is not surprising. And on closer analysis, some of the dangers revealed in the 2002 NIH study of 16,608 women on estrogen therapy (page 93) are not as definitive as originally assumed.[15] When examined according to age groups, for instance, women who received HRT within the first decade after the onset of menopause did not show the increased risk of heart disease found in the overall study. On the other hand, the study has been criticized for underrepresenting the dangers of HRT. Potential participants were screened out of the study if they had some of the risk factors for heart disease, diabetes, stroke, or breast cancer, while a percentage of the actual population being prescribed HRT in the real world has those risk factors. In addition, 40 percent of the women in the study dropped out, usually due to side effects. Increased incidences of cancer and heart disease in the women who had the most adverse initial reactions to HRT were, therefore, not tracked.

The current advice from the NIH and the Food and Drug Administration (FDA) is that women should discuss with their doctor the risks and benefits before embarking on hormone therapy treatment (the standard fallback when science knows it doesn't have the answers). They also suggest that the dosage and duration of therapy should be the least amounts possible for achieving the treatment goals. The tone of caution in these recommendations is well founded. At least 25 percent of the women who do start HRT discontinue it due to side effects. The official recommendations, however, assume that doctors can make good choices based on bad information, with the great error in the original study being that rather than natural progesterone, the synthetic and far-inferior progestin was used.

We don't know from the NIH study what impact natural progesterone might have had. The molecules in natural progesterone fit the receptor sites in a woman's body like lock and key. The molecules in the synthetic progestins have been altered so that they could be patented, and the body often does not respond in the desired ways, with adverse reactions being common.[16]

The late John Lee, M.D., investigated the effects of natural progesterone on thousands of menopausal women. He suggested that many of the symptoms that are treated with estrogen are actually caused by progesterone deficiency, not estrogen deficiency. Too much estrogen, in relation to progesterone, becomes toxic to the body. The truth is, none of us can avoid inadvertent "estrogen replacement therapy" because our environment bathes us in estrogen. The earlier onset of puberty, which came on average at about 14 years a century ago and is now closer to 12 (with a girl's first period at eight or nine not being unusual), has been linked to the hormones indiscriminately used to puff up meat and dairy products, and to the powerful estrogens in the sprays used on grains fed to animals. Our estrogen levels can actually be raised by car exhaust, liquids we drink from plastic containers, and the chemicals found in nail polish, glue, soaps, cosmetics, and paint remover.

To support this thesis, Lee pointed out that for women in nonindustrial cultures, menopausal complaints are minor or unknown. In many agrarian cultures, there isn't even a word for "hot flash," and symptoms such as vaginal dryness, osteoporosis, and mood swings are not associated with the time when a woman stops menstruating. With the introduction of industrialization to Third World countries, resulting in less exercise, greater calorie intake, and greater exposure to external estrogens, dramatic increases in the incidence of menopausal symptoms have occurred. Lee laments that "we have managed to make menopause, a perfectly natural part of a woman's life cycle, into a disease."[17] Implicated here are environmental pollutants, poor diet, unhealthy lifestyle, cultural attitudes, and the incorrect use of synthetic hormones.

Lee's thesis is that few Western women are truly deficient in estrogen. Rather, at menopause, most become deficient in progesterone. When symptoms such as breast tenderness, weight gain, bloating, and mood swings occur in adolescent girls, they are attributed to increases in estrogen; when these same symptoms occur in women during the transition to menopause, they are attributed to "estrogen deficiency." Rather than estrogen deficiency, "estrogen dominance" is the problem, according to Lee. While estrogen is certainly necessary for good health throughout a woman's life, it becomes toxic in quantities that are too high or that are not balanced by sufficient progesterone. Lee acknowledged that some women may need estrogen supplements at menopause (and estrogen deficiency can be confirmed with a salivary hormone-level test). Most estrogen is made in body fat after the ovaries stop producing it, so these women tend to be thin and petite. But for most women, Lee was suggesting that the pharmaceutical and medical establishments have it upside down when it comes to estrogen replacement.

Lee was not alone. In 1991, feminist Germaine Greer impugned the proponents of HRT for not having established that estrogen deficiency is the cause of what they are attempting to treat,[18] a situation that still exists. Sandra Coney's 1994 *The Menopause Industry* argued that women were making disastrous health choices after being subjected to "a demoralizing propaganda campaign aimed at brainwashing them into accepting that they are 'estrogen deficient,' in other words, defective in their normal state."[19] Jerilyn Prior, M.D., professor of endocrinology at the University of British Columbia, points out that it is "backward science" to assume that the symptoms of menopause are tied to decreased estrogen levels, kind of like "calling a headache an aspirin deficiency disease."[20]

Lee summarized: "The estrogen 'deficiency' hypothesis is not supported by the facts of estrogen blood levels, by worldwide ecologic surveys, or by endocrinology experts."[21] His remedies include simple lifestyle choices such as using natural progesterone, adequate exercise, and good diet, including organic meat and dairy products that have not been filled with artificial hormones. Natural progesterone addresses many of the problems caused by the "estrogen dominance" that Lee believed is almost epidemic. Progesterone cream frequently reduces or eliminates symptoms such as hot flashes, night sweats, lowered libido, and "middle-age" weight gain; helps protect against breast cancer; and helps stop and even reverse osteoporosis.

Women to Women, the first medical clinic devoted to health care for women, by women, is now also a source of Internet information and services (www

.womentowomen.com). They recommend that any hormone replacement therapy use bioidentical hormones, which are manufactured to have the same molecular structure as the hormones made by your own body. Synthetic hormones are intentionally different, not for health reasons, but because the pharmaceutical companies can't patent bioidentical structures, even though they are proving to be more effective and less likely to produce side effects. However, while I agree in theory about bioidentical hormones, I have also seen bioidentical hormones cause side effects. This is why I rely so strongly on energy testing, not only for the choice of supplement, but also for when and how much is needed from one time to the next.

Different women metabolize different hormones in different ways, and their hormonal balances are ever-changing. If the supplement, whether bioidentical or synthetic, throws you into hormonal imbalance, it is doing harm. In addition, so-called bioidentical hormones made by compounding pharmacies are not always truly bioidentical, and they are not as carefully regulated by the FDA as those made by pharmaceutical companies. One lot may have a different concentration in its hormonal level from the next.[22] Any drug carries risks! Women to Women's guideline is to get symptom relief at the lowest possible dosage.

Your doctor, however, is more likely to have been informed by the pharmaceutical industry, which promotes a combination of synthetic estrogens and progestins. Such contradictory information from authorities is mirrored not only in HRT, but also in nearly every issue women face at menopause, from keeping our skin soft to keeping our bones strong. Where do we turn amid the conflicting information?

AN ENERGY APPROACH TO THE CHEMISTRY OF MENOPAUSE

Short of moving to a nonindustrialized culture, many basics of diet, exercise, and lifestyle are within your control and are not matters of controversy. Start there. Meanwhile, the energies that are the living presence of the hormonal changes during menopause can be managed directly.

In my practice, I found that helping a woman come to her optimal hormonal balance during perimenopause, menopause, and afterward would generally proceed along three basic steps. More often than you might think, the first step was in itself sufficient. Simply balancing a woman's energies and teaching her how to keep them balanced led to tremendous improvements in how she felt. Many times the woman initially believed she needed estrogen or progesterone, and balancing her energies caused her body to begin producing the hormones she was lacking. But if this was not enough, the second step was to use energy testing to determine which herbs or hormones would be beneficial. The third step was to figure out quantities and timing for using these substances in harmony with her body while minimizing side effects or overdependence.

First Step: Energy Balancing. Before considering more invasive measures, such as taking hormones that your body didn't produce, a first step is to get your system into optimal energetic balance, which in turn balances your chemistry naturally. You learned the Five-Minute Daily Energy Routine in Chapter 2, and that is the ground floor for keeping your energies strong and balanced during menopause. In this chapter, you will be introduced to (1) a special Menopause Module to add to your Daily Energy Routine, designed to promote hormonal balance for women who are entering or are in menopause, (2) specific exercises for increasing your body's production of estrogen, progesterone, and testosterone, along with methods for figuring out which your body needs, and (3) techniques to address specific symptoms of menopause. Energy exercises that correct a specific problem also help harmonize your entire energy system.

Second Step: Energy Testing Vitamins, Herbs, and Hormones. "The healthy body," according to Christiane Northrup, "is equipped to produce all the hormones a woman needs throughout her life."[23] Using an energy approach supports this natural capacity. However, if you are lacking a vital ingredient necessary for health—and it is not easy today to ensure a truly balanced diet attuned to your unique physiological needs—your body has to work hard to compensate. If you have gotten your energies into good balance and symptoms persist, the next step is to evaluate whether you need to take natural hormones or other substances. Lab tests can give clues, but hormone levels fluctuate markedly on a daily and even hourly basis. Because estrogen is so potent and is produced only in minute quantities compared to progesterone and testosterone, "estrogen dominance" (see page 215) is often at play even with relatively low estrogen levels. Thus, many women get prescriptions for estrogen that their bodies do not need and that may actually be harmful. Meanwhile, many over-the-counter remedies are also available. Some may be exactly right for your body. The challenge, of course, is figuring out what you need. This requires educating yourself, finding conventional medical and pharmaceutical information about the choices you are considering, and consulting sources that favor a more natural perspective. The People's Pharmacy (www.peoplespharmacy.com/index.asp) is a superb source for such information. Once you have identified promising vitamins, minerals, herbs, or hormones, energy testing is the best way I know to determine, day by day, whether what you are considering is what you need.

Third Step: Minimize Dependence, Maximize Balance. Natural hormones and botanical products are readily available. Why not just routinely take such supplements? There are reasons for caution when interfering with nature's way. Adding chemicals that are already produced naturally may result in your body producing less of those chemicals. Balances may be thrown off, such as when my overreliance on progesterone caused a permanent thickening of my uterus. What happens, for women who take HRT, to the natural and desirable increase of testosterone, relative to the estrogen and progesterone, that fosters the power of the crone? This is highly complex territory, and it is not surprising that one expert or another advocates virtually every solution imaginable. We are all so different that almost any avenue will work for some women. So-called experts often develop a career by advocating for everyone an approach that works only for some. The menopause industry—on both the pharmaceutical and the natural-healing sides—is a multibillion-dollar enterprise that lacks good road maps. Again, this is where energy testing becomes invaluable. My needs for progesterone and estrogen change on a daily basis, and energy testing gives me a gauge to learn what my body requires and doesn't require during a particular time frame. As you use energy testing, you also begin to notice patterns, so you become less reliant on the energy test. If, for instance, I am feeling a bit hysterical, it is often a sign that my progesterone level has gotten too low, and I may do energy exercises to raise it or may put a drop or two from a natural progesterone liquid capsule onto my wrist. In the rare cases that I feel deadness, depression, or a top-of-the-head headache, it is usually because I have used too much progesterone and my balance of estrogen relative to the progesterone has been thrown off.

Balance is, again, the essential concept and, again, energy testing can be a great gauge. You will also find that once you are regularly balancing your energies, your need for supplemental hormones will decrease. The remainder of this chapter details the first of these three basic strategies, energy balancing. It presents an add-on module to the Daily Energy Routine for women who are dealing with menopause. It shows you how to use energy techniques to increase the production of estrogen, progesterone, thyroid, or testosterone, as needed. And it discusses specific symptoms that are often associated with menopause, such as osteoporosis, hot flashes, and depression, suggesting ways of working with your energies when such symptoms occur.

Adding a Menopause Module to Your Daily Energy Routine

The Daily Energy Routine (page 51) is a cornerstone of how I teach energy medicine. One exercise, however, if you need it, even trumps the daily routine, and that is the Homolateral Crossover (page 67). If you are short on energy, brain power, motivation, or spirit, or if you are depressed, your energies are probably moving in a homolateral pattern. You don't need an energy test to tell you that something is off. You can, however, verify the homolateral pattern with the energy test described on page 68. If your energies are homolateral, start with the Homolateral Crossover, even before the Daily Energy Routine. If your energies are not crossing over, little else you do—exercise, massage, or energy techniques—will yield its full benefit.

If you are in the midst of menopause, or just noticing the first signs of menopause, I suggest that you add the following six techniques to your Daily Energy Routine. As a group they will require less than five additional minutes, and they will have a balancing effect on your hormones. Three of them have already been introduced as part of the Premenstrual Module: Connecting Heaven and Earth (about two minutes, page 44), the Spleen Meridian Flush and Tap (about 30 seconds, page 136), and the Abdominal Stretch (about 30 seconds, page 134). To these, add the Diaphragm Breath, the Hormone Hook-up, and Sunrise/Sunset. Insert all six techniques right after the Lymphatic Massage (page 59). Then end the routine as usual, with the Blow-out/Zip-up/Hook-in.

Diaphragm Breath (time—about 30 seconds)

The diaphragm is a strong, thin shelf of muscle partitioning the chest from the abdomen. It fans oxygen throughout the body. The following exercise not only helps distribute oxygen to every cell, gland, and organ, it also enhances the coordination and effectiveness of the hormone system.

Figure 6-1
DIAPHRAGM BREATH

1. Firmly place your left hand under the center of your rib cage and place your right hand on top of it. With your hands flat, pull your elbows close to your body so you are hugging your midsection (see Figure 6-1).
2. Inhale deeply and push your body toward your hands while your hands push back against your body. Hold your breath and push hard. Although there is no set amount of time, the longer you hold your breath and push (without becoming light-headed), the better.
3. Release your breath naturally, along with your hands.
4. Relax. Do two more times.

Hormone Hook-up (time—40 seconds)

The pineal, pituitary, and hypothalamus glands form a complex axis. Through direct feedback to one another they control the release of many of the hormones that govern your body. Points on the scalp and skin called neurovascular reflex points, when touched lightly, increase the flow of blood in the general area of the point being touched. By holding more than one neurovascular reflex point, you can increase the flow of blood between the areas being held. Because the bloodstream is a primary mode of transportation for the hormones, touching the specific neurovascular reflex points described below coordinates the hormones regulated by the pineal-pituitary-hypothalamus axis, three of the glands that most strongly influence the body's hormones. Doing the Hormone Hook-up daily is like taking an energy vitamin that supports harmony among your hormones. It has the added benefit of reducing stress and helping clear the body of stress chemicals. It feels good and is good to do on a routine basis as well as any time you sense you might benefit from it.

1. Bring the thumb, index, and middle fingers of your right hand together and place this "three-finger cluster" at the top of your head.

2. Simultaneously make a three-finger cluster with your left hand and place it just over the curve behind the top of your head. Hold both points for the length of three deep breaths.

3. Flatten your right hand so your palm is on your forehead and your middle finger is at the point on the top of your head (see Figure 6-2). Relax into this hold for another three deep breaths.

Figure 6-2
HORMONE HOOK-UP

Sunrise/Sunset (time—30 seconds)

The following exercise can help stabilize blood pressure (it lowers high blood pressure and raises low blood pressure) and is generally very good for the heart. In addition, it calms body and mind and tends to bring equilibrium to hormones that are out of balance.

1. Begin with your hands at the sides of your body, palms facing outward. Slowly with an in-breath, circle your arms above your head, as if directing the sun to rise (see Figure 6-3a).

2. With both arms now extended upward and palms facing each other, reach for the risen sun. Stretch one arm up high toward the sun. Imagine grabbing a rope with this hand. Close your fist around it and pull it down, drawing the sunlight onto you. As this hand comes down, reach up to grab the rope with your other hand and pull it down. Keep drawing the sun, alternating hands for several more pulls (see Figure 6-3b).

a   b   c

Figure 6-3
SUNRISE/SUNSET

3. Turn your palms outward. Release your breath slowly as you circle your arms far out to the sides and down to the sides of your legs, as if directing the sun to set and your body to calm (see Figure 6-3c).

Daily Energy Routine with Menopause Module. The entire expanded routine still requires under ten minutes per day once you have learned each technique. Begin with the Homolateral Crossover if required. Then, while the order is not overly important, one good sequence is:

1. Three Thumps (30 seconds, page 53)
2. Cross Crawl (30 seconds, page 56)
3. Wayne Cook Posture (90 seconds, page 56)
4. Crown Pull (30 seconds, page 58)
5. Lymphatic Massage (60 seconds, page 59)
6. Connecting Heaven and Earth (2 minutes, page 44)
7. Abdominal Stretch (30 seconds, page 134)
8. Spleen Meridian Flush and Tap (30 seconds, page 136)
9. Diaphragm Breath (30 seconds, page 220)

10. Hormone Hook-up (30 seconds, page 220)
11. Sunrise/Sunset (30 seconds, page 221)
12. Blow-out/Zip-up/Hook-in (30 seconds, page 74)

ENERGY TECHNIQUES FOR INCREASING THE PRODUCTION OF ESTROGEN, PROGESTERONE, THYROID, AND TESTOSTERONE

The above methods organically help keep your hormones in a better balance. The following methods can be used to increase the production of specific hormones. You can use lab tests to find out if you have hormone deficiencies or excesses, but reliability is a problem due to daily fluctuations. More reliable once you master the art of energy testing is to place a natural form of the hormone in your energy field and energy test it. But even without any tests, certain symptoms are signs of specific imbalances. While only rough indicators (each person's body is different, the same symptom may have numerous causes), and you might want to verify with lab tests, you can also experiment with the energy techniques for increasing production of hormones that seem to be deficient and observe the effects.

Stimulating Estrogen and/or Progesterone Production

The fact that you are no longer wired to bring forth a baby after menopause doesn't mean that estrogen production has stopped, though it is at only one-tenth its previous level (produced now primarily in the adrenal glands and fat cells). Progesterone production, on the other hand, is nearly absent, though tiny amounts are still made in the adrenals and fat cells, and are actually necessary for the continued production of estrogen.

Progesterone production can, however, be inhibited by the hormones found in meats, dairy products, and pesticides. In addition, because everything from soap to nail polish leaks unnatural estrogens and estrogen-mimics into our bodies (which bind them to our estrogen receptors), we may need to balance this estrogen inva-

223

sion by stimulating the natural production of progesterone or by adding it as a supplement through gels or creams. At the same time, the estrogen absorbed from improbable sources such as car exhaust is not the estrogen that wards off osteoporosis. You may need to produce or introduce natural estrogen, as well. Maintaining a balance between estrogen and progesterone is as important a challenge in postmenopause as at any time before.

How do you know if you need more estrogen, more progesterone, neither, or both? I wish I had an easy answer or guideline. The truth is that what you need at one time may be altogether different from what you need at another. These balances are always fluctuating, and even the symptoms of a deficiency or an excess of progesterone or estrogen may be different from one woman to another. I once put the following note on my refrigerator to remind myself of the patterns in my own body: "Dragging or no motivation—estrogen down! Panic or fury, progesterone down!"

These generalizations probably hold for most women, as do those in Table 1. However, they are not an infallible guide. For most women, estrogen deficiency causes hot flashes, and that is the prevailing belief. Yet it is when my own balance goes even slightly toward *too much* estrogen that I get hot flashes. Even from eating tofu, or any other substance with estrogen in it, I get not only hot flashes but also such night sweats that I may wake up drenched. My own experience has led me to a theory that hot flashes are not caused by estrogen deficiency per se. I believe they are caused by rapid shifts in estrogen-progesterone balances. It is also the case that for most women, the imbalance that causes hot flashes is on the side of estrogen deficiency, but for some women the imbalance can be in the other direction.

In addition to individual differences, if a woman takes excess amounts of either hormone, unexpected effects may result. For instance, if you realize you are low on progesterone because you are feeling overly emotional, but then apply so much progesterone cream that you create a relative estrogen deficiency, you may have brought on feelings of depression. Because I periodically use progesterone cream, I have inadvertently done just that. Another way I know I need estrogen is that I get a headache on the top of my head. But if I try to correct it with the amount of estrogen in the lowest-dose bioidentical pill available, I will soon be experiencing the symptoms of progesterone deficiency. Instead, I will do an energy exercise to bring about the balance. This usually works. If it doesn't, I cut the lowest-dose estrogen tablet available to a fraction of its size, and the headache will very quickly be gone. Regardless of how I handle it, I usually do not need external estrogen again for a good while.

Table 1: Possible Symptoms of Progesterone and Estrogen Imbalances

PROGESTERONE DEFICIENCY/ ESTROGEN EXCESS	ESTROGEN DEFICIENCY/ PROGESTERONE EXCESS
Physical Symptoms	**Physical Symptoms**
Breast Swelling/Tenderness	Hot Flashes and/or Night Sweats
Bloating	Insomnia
Water Retention	Vaginal Dryness
Weight Gain	Vaginal Thinning
Fatigue	Uterine Cramping
	Headaches
	Decreased Sexual Response
Emotional/Mental Symptoms	**Emotional/Mental Symptoms**
Anxiety	Sense of Futility
Panic	Depression
Quick Temper	Decreased Caring
Nervousness	Apathy
Withdrawal	Dullness
Avoidance	Mental Fuzziness
Oversensitivity	

The way I suggest you approach this complex territory is that you begin by making some educated guesses regarding your own progesterone and estrogen balances based on the above discussion and any energy testing you have done. Then apply the Daily Energy Routine with the Menopause Module (page 219). This in itself is designed to bring your hormones into a better balance. There is a high probability that you will be feeling better immediately after doing the exercises. If you still sense that you are not in a hormonal comfort zone after doing the Menopause Module for about three days, do one of the following to produce more progesterone or estrogen, based on your best guess of what you need. If you guess wrong, you will not get the relief you are seeking, but no real harm will be done since the body tends to self-regulate when hands-on energy interventions are used.

Preliminaries. Prior to stimulating the production of estrogen or progesterone using energy methods:

1. Help balance the endocrine system with the Triple Warmer Smoothie (page 107).

2. Sedate the liver meridian (page 146), which will prepare the liver to process the hormone increase.

3. Do the Triple Axis Hold (page 134) to optimize the balance in your hormonal system.

4. Then use the appropriate techniques from Table 2.

Table 2: Stimulating Progesterone and Estrogen Production

TO STIMULATE PROGESTERONE PRODUCTION	TO STIMULATE ESTROGEN PRODUCTION
1. Sedate the kidney meridian (see page 146)	1. Massage the adrenal reflex points for two deep breaths.
2. Tap the kidney meridian strengthening points	2. Hold the neurovascular points at the temples and behind the knee for four deep breaths on one side and then on the other.
3. Strengthen the spleen meridian (see page 132)	
Do this routine two or three times per day.	*Do this routine two or three times per day.*

Details on the Specific Procedures in Table 2

To Stimulate Progesterone Production (time—about 10 minutes)

1. After the preliminaries above, sedate the kidney meridian (time—3 minutes, see page 145).

2. Tap the kidney meridian strengthening points (time—about 30 seconds): with a firm tap and a steady rhythm of about one tap per second, tap about two inches above the inside anklebone of each foot for about three deep breaths.

3. Strengthen the spleen meridian (time—6 minutes, see page 132).

To Stimulate Estrogen Production (time—a bit more than a minute)

1. To locate the adrenal reflex points, touch an inch above the navel and then go out an inch to each side (see Figure 6-4). Massage these points firmly for about 15 seconds.

2. With your hand open, lightly place the second, third, and fourth fingers of your left hand on your left temple, just outside the eye socket. Place the index and middle fingers of your right hand in the crease behind your right knee. Relax as you hold this position for four deep breaths. Repeat on the other side. These are the neurovascular points for the triple warmer and gallbladder meridians, respectively, and this combination stimulates the adrenals to produce estrogen. Recall that neurovascular points are always held gently.

Figure 6-4
STIMULATING ESTROGEN
PRODUCTION

Have these methods been scientifically confirmed? They have not. They grow out of my practice and experimentation with thousands of women, as well as my own and their observations about subsequent shifts in mood, energy patterns, and physical symptoms. While they are based on the empirically supported theory that chemistry follows energy, I look forward to laboratory research measuring hormonal changes after the various techniques. This will beyond doubt further refine the methods and also provide better information on how best to match the procedure with the person.

Meanwhile, register how you feel after each technique in the Menopause Module for stimulating progesterone or estrogen. Do you feel calmer? Do you feel like you have your old self back? These are signs that your hormones are in better balance and more effectively reaching their mark. Some women, however, may find that even though these exercises make them feel better, they don't fully counteract the plummeting hormones that may be involved with menopause. In either case, you may still want to use the exercises because they build energetic habits that can help you go through menopause and postmenopause more easily. And you can also consider a range of supplements that may help your body find an optimum adjustment to the changes it is navigating.

Vitamins and Herbs That Enhance Estrogen and Progesterone Balance

I recommend here supplements that I, or my clients, have personally found particularly helpful as they moved into and through menopause. I have seen so many benefits from flaxseed and flaxseed oil that if they were illegal, I would owe it to my clients to become an underworld flaxseed dealer. Beyond its many well-established health-promoting and anticancer properties, **flaxseed** keeps your skin moist, bones stronger, and body juicier through menopause. Interestingly, according to Susan Lark, M.D., the lignans found in flaxseed also help eliminate estrogen deficiency by binding with unoccupied estrogen-receptor sites and mimicking the body's own estrogen.[24] **Essential fatty acids** (omega-3 and omega-6) are also great supplements. They help in the production of female sex hormones, and they also help keep the body moist and supple. **Black cohosh** is an herb that contains phytoestrogens (chemicals that plants produce that act like gentle forms of the estrogen produced in our own bodies), and for some women, it eliminates their hot flashes. I have loved **Mexican yam root** for the gentle form of progesterone it provides. **Ginkgo biloba** helps with memory and other mental capacities as well as circulation. **Ginseng**, a medicinal plant used in China for thousands of years, can counter fatigue, build stamina, and produce estrogen and progesterone, as well as support all the endocrine glands. Dong quai is used for a range of menopausal symptoms, from hot flashes to insomnia.

I also recommend taking a good-quality multiple vitamin and a separate full-spectrum mineral supplement. The processed food we generally consume does not usually supply us with the range of nutrients that we need. However, when it comes to nutrition, herbs, and vitamins, we are all very different, so a great deal of personal research is called for. Use energy testing to the best of your ability to determine in advance what might be helpful and what might not be. Inform yourself as much as possible through books and the Internet about any supplement you are considering. Discuss such choices with health or dietary professionals, and remember that introducing any substance into your body is an experiment, so observe carefully. Be alert for unwanted effects, which may indicate that you are taking too much, mixing too many supplements, or simply taking something your body can't assimilate. And proceed with optimism that the attention you are giving to finding out what your body needs is going to pay off big.

Supplemental Estrogen and Progesterone

I have at my disposal all of the above advice and methods, apply them as needed, and still take and love natural progesterone and, on rare occasions, natural estrogen. Am I a fraud? We live in extraordinary times. We have never been more challenged. And we have never had more resources for meeting our challenges.

The environment strains our natural chemical balances. The many stresses of modern life strain our natural chemical balances. Living longer and in less harmony with nature stresses our natural chemical balances. Energy medicine can do much to restore them. So can the judicious use of the various supplements that are readily available. Better living through energy. Better living through chemistry.

My first advice is to not accept advice from anyone. At least not uncritically. Cookie-cutter recommendations and prescriptions are for indistinguishable cookies. You are unique. But the pharmaceutical industry spends untold millions telling you that moving forward naturally and according to your body's plan is the path to becoming old and decrepit, and that the solution is to take their drugs, which are designed to reverse time and give you the chemistry that you had when you were 25. It is a lie based on a myth. On the other hand, if you take the time to experiment with energy techniques and natural supplements, you can keep your body healthy, supple, and vibrant well into old age.

If you have determined through the various means discussed earlier that your body is low in progesterone and needs more than the methods presented can produce, my suggestion is to begin by experimenting with progesterone creams that the staff at a good health food store recommends (also do a critical search on the Internet). Regardless of the dosages recommended, energy test and experiment.

Recommended dosages are averages based on many women. You need to find the dose needed for *this* woman, and at a particular point in time. If you find that you do not get enough relief, you may need to get a prescription. While your doctor may have other up-to-the-minute suggestions, Prometrium is an excellent form of progesterone, made from yam and encapsulated with peanut oil (and thus contraindicated if you are allergic to peanuts). While it comes in a capsule that is meant to be taken orally, you can instead pierce the capsule with a pin and rub the liquid onto the inside of your wrist or your stomach. This sends the progesterone directly into your bloodstream, bypassing your liver and your entire digestive system, and it also allows you to easily modulate the amount. Another form that bypasses the digestive system is the vaginal suppository available by prescription, though not all pharmacies are able to make it. As discussed earlier (see page 127), it is very easy to tip the scales too far to the other side, so be very alert to the symptoms of too much progesterone, described on the right side of Table 1 (see page 225).

Your body needs a balance of progesterone and estrogen to support your heart, keep your bones strong, and help your brain stay sharp. If you have determined that you need supplemental estrogen, the choices include phytoestrogens, herbal compounds, bioidentical estrogen, and standard pharmaceutical hormone replacement therapies such as Premarin. The same principles apply in terms of informing yourself; energy testing to the extent of your abilities, starting with the more natural compounds (though I have sometimes even found standard pharmaceuticals to energy test stronger than bioidenticals—again, individual differences trump everyone's theories); and considering taking less than the recommended dosages. For instance, the lowest-dose Premarin tablet can be cut in half and cut in half again with a pill cutter, and even a third time, and meets the needs of many women. Premarin is also generally taken on a daily basis, yet unlike an antibiotic, there is no compelling logic for this if you are able to carefully monitor your day-by-day needs using self-observation or energy testing. With estrogen, it is also now standard treatment to supplement it with progesterone, as the harmful effects of "unopposed" estrogen replacement therapy are well documented. In any case, after taking a hormonal supplement, tap the spleen neurolymphatic reflex points on both sides of the body (page 136) to help you metabolize it.

Thyroid

Until 2002, all thyroid tests were developed and interpreted on the basis of lab data from men, even though nearly 90 percent of the complaints doctors receive that might involve thyroid dysfunction are made by women. Because of this little oversight, many women with symptoms of thyroid dysfunction were showing up as normal on lab tests and not receiving needed treatment. The problem was so extensive that there was actually a "hypothyroid patient movement," supported by knowledgeable endocrinologists, demanding that tests be made more sensitive and that supplemental thyroid be more readily prescribed.

Meanwhile, others have decried this movement, not because they object to symptoms being alleviated, but because supplemental thyroid hormones are, according to Diana Schwarzbein, M.D., "Band-Aids to cover up symptoms. They do nothing to restore balance to your body." She counsels: "If you take thyroid hormone replacement therapy for a lifestyle-based endocrine disorder instead of improving your nutrition and lifestyle habits, you will actually create more hormonal imbalances and further destroy your metabolism."[25] She insists that proper nutrition is the first order of business. Since hormones are made primarily from proteins, cholesterol, and essential fats, a balanced diet is critical for maintaining the production of your hormones. Adequate amounts of iodine are required. Herbs that can help stimulate your own production of thyroid hormones include nettle leaf, kelp, parsley, mullein, selenium, and Irish moss. A full-spectrum mineral supplement is generally valuable and is specifically helpful for the thyroid. Progesterone also supports the thyroid.

Schwarzbein points out that because the body makes hormones in minuscule and fluctuating amounts, it is very hard to answer many hormonal questions using blood or urine tests. While I cannot advise you on whether to take thyroid replacement therapy, using energy medicine to help balance your thyroid is a noninvasive approach, more akin to nutrition and lifestyle interventions, and it may prevent the need for more extensive measures. As we age, our thyroid gland tends to produce fewer of the thyroid hormones that are essential for metabolism, vitality, and good mood. Symptoms and signs that your thyroid hormone production is off are shown in Table 3.

Table 3: Some of the Major Symptoms and Signs of Thyroid Imbalance

THYROID DEFICIENCY (Hypothyroidism)	THYROID EXCESS (Hyperthyroidism)
Feeling Cold When Others Don't	Warm and Moist Hands
Low Body Temperature	Heart Racing and Pounding
Slow Pulse, Low Energy Level	Huge Appetite, no Weight Gain
Extremely Dry or Flaky Skin	Exhaustion
High Cholesterol	Bulging Eyes
Brittle Fingernails, Hair Loss	Easy Bruising
Can't Lose Weight, Muscle Weakness	Insomnia
Unexplained Sadness, Brain Fog	Tremors
Trouble Swallowing or Breathing*	Graves' Disease
Hoarse Throat and Coughing*	Hair Becomes Softer and Finer
Enlarged Lymph Nodes*	

*These symptoms may be signs of a serious condition and should be medically evaluated.

Physical exercise and adequate sleep are essential for proper thyroid function. Simply stretching the skin above, below, and to the sides of the Adam's apple instantly brings good energy to the thyroid. In addition, the following techniques, alone or in combination, reduce stress in the thyroid and stimulate the production of thyroid hormones.

Figure 6-5
Thyroid Booster

Thyroid Booster
(time—about a minute)

1. Do the Triple Warmer Smoothie (page 107) and then tap the spleen neurolymphatic reflex points (page 136).
2. Place the middle finger of one hand just above the Adam's apple, the middle finger of the other hand just below the Adam's apple, and separate your fingers so you are stretching the skin

on your neck. Do the same with your fingers on the sides of your Adam's apple. Also stretch diagonally.

3. Rest the thumb, index finger, and middle finger of one hand in the hollow below your Adam's apple and rest the pads of the fingers of your other hand at your temple (see Figure 6-5). Both are triple warmer neurovascular reflex points. Hold gently for four or five deep breaths and then repeat on the other side.

Testosterone

Women produce about one-seventh the testosterone that men do, primarily through their ovaries and adrenal glands. The production of testosterone declines substantially just prior to menopause, at which time a woman produces about half the amount she was producing in her midtwenties. However, after menopause, the ovaries continue to make testosterone in quantities such that testosterone is proportionately higher than estrogen as compared with earlier in life, supporting some of the increased assertiveness and independence seen in postmenopausal women.

The doctor who gave me my annual exams in the early 1980s was a feminist who said to me one day: "You know, Donna, you'd like your life a lot better if you weren't *so* nice. A lot of women are experimenting with testosterone, and they are loving it. I think it would be really good for you. You'd strut around owning your space, and you'd have clear 'no's' for people." She talked me into trying it. She had some samples in her office, and she gave me the lowest dose available. She hoped I would take it and tell her what I experienced. It did make my life different. I was furious all the time. And not the kind of juicy anger that engages you with the people you love; it was more like a marine about to attack an enemy squadron that had just wiped out an innocent village. While I stopped taking it almost immediately, it did give me much greater respect for men who are naturally juiced up with that stuff all the time and are still able to behave in a civilized manner.

So I am not going to offer suggestions about taking testosterone replacement—you can weigh the risks—plus there are other forms besides the synthetic macho type that I took. **Muira puama**, for instance, is a plant from Brazil whose bark and roots have been shown to increase libido in menopausal women by upping testosterone levels. **Maca** is a plant from Peru whose root has a similar effect. **Damiana** is a shrub, also from South America, that has been used as an aphrodisiac since Mayan times. While I do not have substantial experience with any of these, I can offer some energy interventions if you think your testosterone levels may be lacking. Table 4 describes symptoms and signs of testosterone imbalances.

In the early 1980s, I did volunteer work at a women's shelter. These women had been abused and had come to the shelter for safety and refuge. They were so deeply exhausted and demoralized that they were terrified of leaving, even if they didn't particularly like being there. I remember one little boy who saw me giving a session to his mother. She began to sob as she relaxed on my table, and he must have thought I was hurting her like his dad had hurt her. He started beating on my leg with all the fury his little three-year-old body would allow. It was actually quite poignant. Testosterone in action! His mom, however, seemed totally deficient of anything like such a life force, as were most of the women when they got to the shelter. No fight was left in them. They had no reserve power. We all have yin and yang, female and male energies within us, and we need each. These women were depleted of both. Triple warmer, which governs the fight/flight/freeze responses, had been on continual alert for a long time, and they had used all they had to fight and flee their way to the shelter. Now their energies were in freeze mode. I'm sure their testosterone levels and corresponding yang energies were as low as at any time in their lives. Yet they needed those yang energies if they were going to survive in the world and not return to their abusive partners. So I began sedating triple warmer meridian and strengthening its counterpart, spleen meridian. You might think that sedating triple warmer would take away power, but it had the opposite effect. I would calm it and calm it and calm it, and I could see the yin forces in the body gathering, gaining strength. Then the yang forces. In the process, testosterone production gets back on track.

The other meridian involved in producing testosterone, in addition to triple warmer, is the stomach meridian. While a yang or "male" energy, the stomach meridian in a woman carries the force of the mother tiger. During her mothering years, this energy is focused on children and family. After menopause, it is available for establishing the power of the crone. It is a grounding force that helps a woman hold her own space. So to set the stage for more testosterone production, you want to be sure your stomach meridian is strong, steady, and uncompromised. Steps include regularly doing the Stomach Meridian Flush (page 138). Precede it with the Blow-out (page 74), the Zip-up (page 61, with an affirmation about stepping into

Table 4: Some of the Major Symptoms and Signs of Testosterone Imbalance

TESTOSTERONE DEFICIENCY	TESTOSTERONE EXCESS
Feeling Powerless	Increased Anger and Assertiveness
Decreased Libido	Deepened Voice
Decreased Energy	Increased Facial Hair

the power of the crone), and the Three Thumps (page 53). Then connect the energies between the triple warmer and stomach meridians by placing your thumbs on your temples and your fingers directly above your eyes. These points get the energies of both meridians in sync with each other. Finally, having created a readiness in your body to utilize more testosterone, the Testosterone Triple Massage is designed to activate testosterone production. While I have never evaluated this sequence with lab tests, women have reported that it gave them more power, stronger boundaries, greater mental clarity, emotional balance, and renewed libido. To naturally increase testosterone, do the Testosterone Triple Massage two or three times per day:

Testosterone Triple Massage
(time—less than a minute)

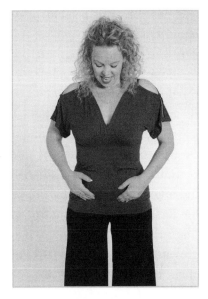

1. Place your middle fingers in your navel, move them up one inch, and then out one inch to either side. Take a deep breath and massage these points deeply for about 15 seconds. This stimulates the adrenal points, strengthening the adrenals and enhancing testosterone production.

2. Place your middle fingers at the knob on the top inside edges of your hip bone. Move your fingers two inches toward the center, breathe in deeply, and massage for about 15 seconds (see Figure 6-6). This stimulates and strengthens the ovaries and enhances testosterone production.

Figure 6-6
TESTOSTERONE TRIPLE
MASSAGE

3. Deeply massage the points on the outsides of your legs between your knees and hips for about 15 seconds (to find these points, hang your arms and bend your middle fingers toward your legs). These are gallbladder meridian points that stimulate the yang forces in the body.

Osteoporosis

A woman who had broken a rib by simply sneezing sought me out as a last resort. The only indicated medication at the time for reversing bone loss was causing severe side effects. She was only 49, and she did not want to see her body deteriorate further. But she was also a physician, well educated in Western medicine, and did not believe she could reverse her condition, citing research in our opening minutes together to support this view. She hoped instead that if she could build the muscles supporting her skeletal structure by using physical exercise and could strengthen the energies in her body (she knew very little about energy work, but I had helped a few of her patients and she was open to giving it a try), she would be less vulnerable to the effects of her "genetically destined" loss of bone density. I felt that more was possible, but I decided to start the session by showing her how energy work could both help lift her sense of despair and leave her feeling better in her body. This gained me enough credibility that I could talk to her about how energy feeds the bones and can literally reverse bone loss. She was skeptical but willing to let me try. Bone density is regulated by the kidney meridian. But for her, a simple formulistic approach was not enough. The kidney meridian is fed by the lung meridian, and it was a deficiency in her lung meridian that was keeping her kidney meridian weak. So we wound up giving as much attention to her lung meridian as to her kidney meridian. The lung meridian governs grief, and to her surprise, one of the keys in helping get her lung meridian back to an optimal flow involved emotional work around a loss she had not fully processed. After four or five sessions and faithful compliance with back-home assignments on her part, we both knew her bones were growing stronger. After a bone density test showed dramatic improvement, she spoke to me about doing a research project, one of those opportunities I never quite had time for. But it is a project I hope someone reading this book will carry forth.

Osteoporosis afflicts more women than stroke, diabetes, breast cancer, or arthritis. It is a substantial challenge that causes a progressive loss of minerals, bone mass, and bone density. It may have increased as much as sixfold in recent decades.[26] It is also surrounded by misinformation. Estrogen, for instance—whose decrease during menopause has been identified as the culprit in osteoporosis—plays a far less significant role in protecting against the disease than does progesterone.[27]

Your bones are in a constant process of breakdown and regeneration, a miracle of renewal that depends on two kinds of cells. The first are called osteoclasts, which are modulated by estrogen, and take away old, worn-out bone. The second are called osteoblasts, which are modulated by progesterone. They go to those same sites and create new bone. It is critical for keeping your bones strong to maintain an adequate progesterone level.

As we age, osteoclast activity becomes more prominent, and bone mass is lost. But women in their 70s and 80s can still slow their bone loss, strengthen their bones, and even build new bone. Long before that, however, by their 40s or 50s, they can be taking steps to stop the loss of minerals in their bones, particularly calcium and magnesium. Getting these minerals into your system, however, can be tricky.

Calcium pills, widely touted as a preventive for osteoporosis, may interfere with the metabolism of magnesium and zinc, thus ultimately harming rather than promoting bone strength. Calcium is best obtained from your diet and, even then, cannot be adequately absorbed unless your body has the proper amounts of magnesium, hydrochloric acid, and various vitamins. Magnesium may, in fact, be as important as calcium in preventing osteoporosis. Meanwhile, many prescription medications, as well as over-the-counter medications such as antacids, disrupt calcium absorption. The calcium that is naturally found in herbs, such as nettles, is an excellent way to supplement your calcium intake. If you take calcium in a pill form, it is most likely to be absorbed if it is part of a multimineral that contains more magnesium than calcium.

An enormously unfair contributor to osteoporosis is that our bones absorb toxic lead from the environment, which displaces calcium and results in bone loss. Adding further injury, at menopause and after (and also, alarmingly, during pregnancy and breast-feeding), lead in the bones is released into the bloodstream, where its poisonous effects may contribute to hypertension, kidney disease, dementia, heart disease, and nerve damage. Because lead can leach into the bloodstream whenever calcium is released from the bones, one way to keep lead out of your bloodstream is to be sure you have adequate amounts of calcium available. If the body isn't pulling calcium from your bones, it won't be leaking lead into your bloodstream. In addition, several supplements have been documented for helping rid the body of lead. **NAC (N-acetyl cysteine)** enhances urinary excretion of heavy metals, including lead. **SAM-e (S-adenosylmethionine)** reduces the toxic effects of lead and binds lead to the bile for elimination. **Epsom salts (magnesium sulfate)** also reduce lead toxicity. Other nutrients that help with bone strength include **vitamin K,** chondroitin, MSM, and vitamin D (particularly D$_3$). **Vitamin D** is provided most naturally by the sun, but those of us who live most of our lives indoors are often deficient in D. It is also blocked every time we use sunscreen. Don't ever let yourself get a sunburn, but do let your skin soak the sun's rays in moderation.

Simplistic solutions are, however, particularly misleading with osteoporosis. Proper diet, regular *weight-bearing* exercise, and maintaining a healthy estrogen/progesterone balance form a triangle of prevention and bone regeneration. Energy exercises can strengthen the foundation of that triangle. Just as magnets and electromagnetic therapies have been shown to speed the healing of broken bones, subtle energy work as taught in this book helps maintain bone strength and density.

An excellent and very simple exercise that engages your energies to help prevent osteoporosis is the Shoulder Lift. It lubricates the shoulders, releases lymph, helps your upper back stay strong and straight, and even helps prevent dowager's hump. It is valuable not only in the physical results it brings, it also helps you establish a more conscious and positive relationship with your shoulders and upper back. Lift your shoulders to your ears and slowly roll them backward, around, and up again several times while breathing deeply. Then roll them forward, either simultaneously or one shoulder at a time.

Isometrics for Preventing Osteoporosis

Isometric exercise is a type of strength training that has proven to be extremely valuable in preventing and correcting osteoporosis. Loss of bone density had become a problem for astronauts, who might spend months with little physical movement. Just ten minutes of isometric exercises each day can reverse bone loss. Isometric exercise, which involves the static contraction of a muscle without any visible movement, is also great for your energies. Getting some instruction or a book on isometrics is a superb way to address or prevent bone loss. Here are two of my favorite isometric exercises:

Torso Stretch (time—about 20 seconds)

1. Sit tall on your hands with your fingers facing inward and straighten your arms.
2. Breathe deeply while holding this position for several breaths.

You will feel the stretch in your back, neck, and arms.

Bone Saver (time—less than 2 minutes; also a great tummy trimmer)

1. Place your arms out in front of you, take a deep breath in, hold your breath, and push the heels of your hands hard against each other (see Figure 6-7a).
2. At the same time, squeeze your stomach in toward your spine, tighten your buttocks, and tilt your pelvis forward and up.

a   b

Figure 6-7
BONE SAVER

3. When you want to exhale, instead take three more rapid breaths in through your nose while holding the above posture.

4. Then exhale slowly, as if blowing through a straw (your stomach will tighten naturally).

5. When it feels like your exhalation is complete, blow three more rapid breaths out through your mouth.

6. Relax after each round. Do several rounds.

7. Repeat the entire sequence, except this time bend the fingers of each hand, hook them into one another, and try to pull your arms apart so there is isometrc tension at your fingers and in your arms and shoulders (see Figure 6-7b).

In my practice, when someone's bones were weak or fragile, giving them the homework assignment of sedating and strengthening the kidney meridian two or three times per day began to rebuild bone strength in ways that the person could readily recognize within two or three weeks. When my father had bone cancer, he was told that his fifth lumbar vertebra (L-5, a spinal vertebra located at the waist) was "hanging on by a thread." He had to wear a very uncomfortable contraption to keep it stable because if he were to so much as sneeze, they said, the bone could snap and he would be paralyzed for life. One of my frequent extended visits came at about the time he was fitted for the contraption. With a father's unconditional confidence in his daughter's magical powers, he said, "Donna, you can fix this!" I didn't have any inkling that I could fix it, but I applied the tools I had available. I tested all his meridians several times each day to look for patterns. The kidney meridian kept showing weak.

My father's original diagnosis had been prostate cancer. The prostate is governed by the kidney meridian, and the kidney meridian also governs bone strength and the building of bone. I will take a little detour here, as this was one of the experiences that has made me so insistent in advising that you not squash your own judgment in deference to medical authorities. My father's PSA counts had been somewhat elevated. His doctor was not particularly concerned. But my father loved life and decided to get a second opinion from an oncologist. The oncologist insisted that he have surgery, and as the situation unfolded, it was clear that the two physicians strongly disagreed with each other. Fear won, the surgery was performed, and bone cancer was diagnosed soon after. In reviewing the course of the illness, my father's primary care physician was furious. He advised us to sue the oncologist, explaining that during the surgery, cancer cells had "spilled" into his bloodstream and then gone into his bones. He told our family, after my father died, that the bungled surgery had been a "death sentence."

It was. But we did not know that at the time, and Daddy had asked me to fix his back. Since the kidney meridian governs both the prostate and bone strength and was the meridian that was consistently weak, it was my focus. Every day, three times each day, I would sedate and then strengthen his kidney meridian. This was very reinforcing because immediately following each mini-session, he would report that the pain had diminished. Within three weeks, he was feeling no pain in the area of his lower back. He persuaded his doctor to do another X-ray of the area. L-5 was now *normal*. It had grown back completely. This is not supposed to happen! The medical team was so flummoxed that the first thing they did was check to be sure that the original X-ray was not that of another patient. With the kind of cancer my father had, they expected to see only further deterioration, yet the uncomfortable back brace went into permanent storage.

If you have osteoporosis or are concerned about the condition, you want to be sure to keep your kidney meridian strong. You might want to get an assessment from an energy medicine practitioner (www.innersource.net/links/links_practitioners.htm) or from a good acupuncturist. The energy test required is beyond the instructions given in the Appendix. If the kidney meridian is weak, you can sedate it (page 145), which relaxes the kidneys themselves, allowing toxins to move more easily through them. Then follow by tapping the kidney meridian strengthening points (page 226) to infuse it with more power to help build new bone cells. If you do this at least twice per day for a month, and then go in for another bone density test, you are likely to see improvement. Sedating and strengthening the kidney meridian does no harm, even if the meridian is already strong, and in fact is always valuable, kind of like exercising a muscle.

STIFFNESS AND JOINT PAIN

Osteoporosis, in its early stages, is a "silent" disease. Joint pain is not. It screams at you. Many postmenopausal women suffer with joint pain, stiffness, or aches in the back, knees, shoulders, hips, and neck—both on a temporary and chronic basis. Despite decades of wear and tear on our joints, and even injuries, we can counter and help prevent such pain and stiffness by keeping our joints lubricated.

Inflammation is a primary culprit in joint problems. Inflammation is the normal and wholesome process by which the body protects itself from irritation, injury, or foreign substances such as bacteria and viruses, sending chemicals and white blood cells to affected areas. But with unceasing stress, inadequate or excess exercise, or poor diet, inflammation may be triggered on a false-alarm basis, so it is produced when it is not needed and becomes harmful. High cortisol levels are endemic for many people experiencing the stresses of modern life, and sustained high levels of cortisol damage bone tissue and impair its regeneration. While inflammation is the common denominator in numerous complex diseases, from arthritis to diabetes to cancer, it has the straightforward physical effect of interfering in the exquisite dance at the joints where bones meet.

To keep your joints well lubricated, a simple and very basic step you can take is to be sure you have enough essential fatty acids in your diet. Don't let the "fat" in the name put you off. You need them, and your body cannot produce them. You have to get them from what you eat. Unless you live in a fishing village, however, you probably aren't getting enough essential fatty acids if you aren't conscious about your need for them. Fortunately, foods with both types of essential fatty acids—omega-3 and omega-6—are readily available. Natural sources include fish and shellfish, flaxseed and flaxseed oil, hemp oil, soy oil, canola oil, evening primrose oil, chia seeds, pumpkin seeds, sunflower seeds, leafy vegetables, avocados, and walnuts. The health benefits of taking adequate amounts of essential fatty acids are enormous, ranging from preventing inflammation to promoting cardiovascular health. Flaxseed and flaxseed oil are great sources of essential fatty acids, and their multiple health benefits include preventing cancer, helping dry eyes, and keeping the body's tissues moist. You can buy flaxseed oil, mixed with peppers and garlic, as a delicious salad dressing, or make your own. I add lemons and basil leaf as well (which is also good for keeping your body alkaline as you age). But you can also take it in capsules. You can energy test dosage.

Another straightforward way of keeping your joints strong and healthy involves keeping the energies moving freely through them. A number of stretching exercises and energy techniques can help ensure this. Connecting Heaven and Earth (page 44) is excellent. Three more include:

Circular Joint Rub (time—about 30 seconds)

Simply rub, in a counterclockwise circle, the joint pain with the heel of your hand or with your fingers.

Tape the Magnet on the Pain (time—10 to 20 minutes)

Tape the north side of a weak flat magnet (Radio Shack's round donut magnets work well) against the skin over the bone or joint that hurts for about ten minutes. You can use a compass to determine which is the north side. The needle that points to north will point to the side of the magnet that goes on your skin. If the magnet hasn't made a significant difference in the first ten minutes, remove it. If it has made a difference, keep it on as long as you feel it is continuing to work. Do not leave it on overnight. Do not tape the south side against an inflamed joint, and do not place either side over a vein.

Wrist Twist (time—less than a minute)

Stretching creates space that helps lymph and lubricating fluids move through the joints. When trapped energy is freed, it can become a positive healing force. While yoga and other stretching exercises are great to do on a regular basis, the Wrist Twist is specifically for stiffness in the hands, shoulders, and arms, and it is a method you can use in the moment.

Figure 6-8
WRIST TWIST

1. Let your arms hang and then lift them out to the side about 30 degrees.

2. With your hands open and your fingers leading, rotate your hands inward until you feel a stretch in your forearms and shoulders (see Figure 6-8). Hold that stretch for about ten to fifteen seconds.

3. Relax. Repeat several times.

OTHER SYMPTOMS ASSOCIATED WITH MENOPAUSE

The remainder of the chapter presents energy interventions that may be helpful in working with other common challenges women face during and after menopause. Because the same symptom may have a totally different cause for one woman than for another, no list of complaints and remedies is going to be completely reliable. But I can describe the procedures that women in my practice found most helpful for each of the following conditions. If several methods are presented for a condition that is of concern to you, experiment to find which, or which combination, is most effective.

Hot Flashes

In addition to the following energy methods and the Menopause Module, several herbal remedies are widely reported to help with hot flashes. Those that seem to have been most effective with my clients include flaxseed or flaxseed oil, evening primrose oil, and for some women, black cohosh. Many of my clients stopped HRT with no adverse effects using energy exercises with flaxseed oil as the only supplement. If the flaxseed and the Menopause Module are not enough, additional energy methods that are useful with hot flashes include:

The "Cool-Down" Acupoint. While in a hot flash, press the middle fingers of each hand into the point about two inches below your navel (this is the point of intersection of the central and kidney meridians). Take a deep breath and hold the point as you then breathe normally. This quickly redistributes energy in a manner that reduces body temperature during a hot flash.

Triple Warmer Smoothie. Because triple warmer manages body temperature as well as the stress and immune responses, calming it with a method such as the Triple Warmer Smoothie often provides instant relief (see page 107).

The Darth Vader Breath. Another technique that can quickly calm triple warmer and may provide quick relief during a hot flash is the Darth Vader Breath (page 152).

Dowsing the Fires. A fourth technique not only provides relief, it also prevents future hot flashes. I found that when women did the following procedure on a daily basis, their hot flashes became less intense and, in some cases, subsided altogether. This technique is also good for maintaining thyroid health.

1. Cross your hands, placing your middle fingers on K-27 (page 53) and massage both K-27 points with some pressure.
2. Remove your right hand and slide the middle finger of your left hand above the corner of your clavicle.
3. Turning your head to the left as far as you can (see Figure 6-7), slide the middle finger of your left hand out toward the outer edge of your clavicle while you breathe out.
4. Repeat with your right hand on your left clavicle.

Figure 6-7
DOWSING THE FIRES

Depression

For people who have suffered with depression, the deadness that may be experienced during menopause is at least familiar territory. However, if you are prone more toward hysteria, like me, and less toward depression, it can be new and terrifying. I didn't know what to do with the unfamiliar sense of emptiness—the strange lack of feeling, motivation, or caring. Though I always felt compassion, my understanding for people who experience depression increased profoundly by going through it myself. Depression is not only in the mind. Every cell of the body, every organ, every body function becomes sluggish. Your reflexes and thinking are slowed. Too little estrogen in relation to progesterone can cause depression. The Daily Energy Routine with the Menopause Module, along with the other steps already discussed for bringing estrogen and progesterone into balance, will help ward off menopausal depression. Following are more targeted steps. When you are depressed, it is hard to muster the energy to turn the depression around, and you lose incentive. So be alert to the fact that you probably won't feel like doing these exercises, but when you are through, you are likely to know that something has begun to shift.

Stretch, Stretch, Stretch (time—about 3 minutes)

If you are depressed, energy is not moving. You can instantly activate the energies in your body by simply stretching the skin on your face, scalp, and neck. In the process you will be stimulating the endpoints of your stomach, small intestine, large intestine, bladder, gallbladder, triple warmer, central, and governing meridians. Breathe deeply throughout all these stretches.

1. Begin with the Crown Pull (page 58). Then give yourself a scalp massage.
2. Continue to stretch every part of your face.
3. Push in at your cheekbones and move your fingers to the opening of your ears, stretching your skin.
4. Place your middle fingers above your upper lip, push in, and stretch the skin to the outside of your mouth.
5. Repeat, this time starting below your bottom lip.
6. Place your fingers on your jawbone and push up along the line of your jaw.
7. Pull your ears, starting at the bottom lobe and traveling up about half an inch at a time, until you have stretched the entire outer lobes.
8. With your head bent back, stretch the skin on your neck in every way you can.

9. Place the fingers of one hand on the opposite shoulder, push in, and pull forward. Repeat on the other side.

10. Experiment with stretching any other area of skin that might feel good or doing yoga-like stretches of your muscles and joints.

11. Optional power boost: Tap vigorously with several fingers on all the bones throughout your face and head.

Homolateral Crossover (time—3 to 5 minutes)

Any time you are depressed, your energies are running in a homolateral pattern. Period! Do all you can do to get them crossing over (page 65), particularly figure-eight patterns. The Homolateral Crossover (page 67) is also one of the most powerful methods you can use, though you may need to do it two or three times per day for a while to shift the energetic pattern and get the new pattern to stick. Then other methods for countering depression are more likely to be effective. Getting your energies back into a crossover pattern also helps bring your hormones out of the chaos that homolateral energies create.

Mellow Mudra (time—3 minutes)

By combining the main neurovascular points with a hand posture called a mudra in yoga, you are able to stimulate the flow of blood to your head while calming and aligning energies in the body that are reacting against one another, a property of depression. Moving the blood also gets energies moving, and moving energy is the antithesis of depression. Do another Crown Pull (page 58), followed by rubbing your head all over, giving yourself a pleasant massage. Then put your thumbs over the nails of the index fingers of each hand (see Figure 6-10a). Keeping this position or mudra, place your third and fourth fingers gently on your forehead, directly above your eyes and equidistant between your eyebrows and your hairline. Rest the thumb and index finger circles gently on the temples and, with a slight pull to the sides from your third and fourth fingers, begin to hum. Hold these points for two to three minutes as you continue to hum (see Figure 6-10b). Breathe deeply.

a   b

Figure 6-10
MELLOW MUDRA

Heart-Womb Connection (time—3 minutes)

Standing, sitting, or lying down, place one hand over your heart chakra (middle of your chest) and the other over your womb chakra (between your navel and your pu-

bic bone—see Figure 6-11). Notice the energies between these two chakras beginning to connect as you stay in this position for about three minutes, breathing deeply. People have told me that feeling the energies moving between these places attunes them to an aliveness within that counteracts depression.

Mental Sharpness

The same techniques that help overcome depression also help in overcoming the diminished mental acuity that is sometimes experienced by women going through menopause. The Daily Energy Routine, the Menopause Module, and the

Figure 6-11
HEART-WOMB CONNECTION

stretching, crossover, and neurovascular techniques described above are particularly effective in keeping your mind sharp. **Ginkgo biloba** is proving to have brain-enhancing properties. While early research led to mixed findings, more recent studies, based on higher dosages, showed substantial improvement for a range of mental functions, including reversing some symptoms of Alzheimer's. Many of my clients have reported that it helped them with memory and mental clarity. Established biological effects of ginkgo biloba are that it improves blood flow to most tissues and organs, protects against cell damage from free radicals, and blocks many of the effects of blood clotting associated with various cardiovascular, renal, respiratory, and central nervous system disorders. It is contraindicated for certain individuals with blood-circulation disorders. In summary, the exercises mentioned above, possibly combined with brain-enhancing supplements such as ginkgo biloba, can help your mind stay vibrant and sharp.

Spongy Tummy and Thickening Waist, Oh My

Think oxygen. While Chapter 7 is all about weight management, start by increasing your oxygen levels. A few facts:

- Oxygen breaks down fat molecules, turning them into carbon dioxide (CO_2) and water (H_2O).
- Your body is continually processing toxins—your health depends on it—and you are designed to discharge, with your breathing, more than two-thirds of the toxins your body processes.
- If oxygen is in short supply, the ability of the villi in the intestinal tract to absorb nutrients and digest food effectively plunges by as much as 72 percent.
- When people use only the top part of their lungs to breathe, their metabolism is slowed and other problems ensue.
- The metabolic rate can instantly increase by up to 30 percent when oxygen intake is increased.
- Deep breathing exercises can triple the fat-burning oxygen available to us and increase by up to 70 percent the digestive system's ability to convert toxins into gases, which are then exhaled.

Research has shown that introducing three five-minute segments of daily breathing exercises can substantially increase metabolic rate. It also improves muscular strength, toning, and emotional outlook. Diaphragm Breath (page 220), part of the Menopause Module, is a 30-second daily breathing exercise designed to remind your body how to breathe well. You can supplement it with the following isometric exercise, which I call the Metabolic Breath, or the Menopausal Tummy Trimmer. The basic principle is to breathe deeply in the manner described while tensing several muscles.

Metabolic Breath (time—about a minute, with a nod of appreciation to Greer Childers)

1. Bend forward at the waist and place your hands on your knees.
2. Take a fast, deep in-breath through your nose. As you do, round your back and squeeze your stomach in toward your spine.
3. Force your breath out through your mouth (see Figure 6-12a).

a   b

Figure 6-12
METABOLIC BREATH

4. When all your breath is out, suck your stomach in tight and hold without breathing until you want to catch your breath (see Figure 6-12b).
5. Release. Your lungs will fill with air.

VAGINAL DRYNESS

Women I've worked with who complained of vaginal dryness reported substantial improvement after increasing their *daily* intake of essential fatty acids. Again, sources include fish oils, flaxseed and flaxseed oil, hemp oil, soy oil, canola oil, evening primrose oil, chia seeds, pumpkin seeds, sunflower seeds, leafy vegetables, walnuts, and avocados. While avocados are always included on the lists of good oils to take, I can give them a personal plug from my own experience. When I moved to an area where avocados were plentiful, I noticed within a couple of months that my skin and rough feet became markedly softer, moister, and suppler, even though my new home was in a drier climate. In figuring out why this might be after years of leathery feet, I realized that the only real change was a substantial increase in my intake of avocados, usually as guacamole (well, maybe it could have been the margaritas I had *with* the guacamole?). The Menopause Module also helps promote moist skin and prevent vaginal dryness. Occasionally, additional energy work is needed, and this focuses on keeping the kidney meridian balanced by sedating it (see page 145) and then tapping its strengthening points (see page 227).

HEADACHES

Headaches that come with menopause may have many causes. Your energies may be off. Your life may be full of stress. Your sleep patterns may be disturbed. Your hormones may be out of balance. I found that yam roots could help me since I tended, like many women, to be low in progesterone. For other women, bioidentical estrogen might restore the balance. Exercises that free energy that gets trapped in the skull can sometimes provide immediate relief. Start with a vigorous Crown Pull (page 58). That may be enough. If not, the Pendulum Head Swing is an excellent next step.

Pendulum Head Swing (time—about two minutes)

Figure 6-13
PENDULUM HEAD SWING

1. Take a deep in-breath.
2. Letting your breath out, drop your head forward to your chest, and relax your shoulders.
3. Breathe in as you let your right ear pull your head over your right shoulder and feel the stretch on the left side of your neck (Figure 6-13). Take two or three deep breaths in this position.
4. Slowly as you let your breath out, let your head complete the circle, with your left ear leading toward your left shoulder. Feel the stretch on the right side of your neck as you take two or three deep breaths.
5. Come back to center, dropping your head forward to your chest.
6. Repeat at least two more times.

INSOMNIA

While insomnia may have many causes, physiological as well as psychological, calming the overactive or chaotic energies that contribute to insomnia often works at a deeper level and brings peaceful sleep. You may have to experiment with a variety of methods and find which works best for you. Beyond energy methods, sleep labs (www.sleepcenters.org) have studied insomnia and innovated various techniques you might want to investigate. They are described in a number of popular books as well as various Web sites. One of the most common causes of insomnia for menopausal women is low magnesium levels. You can energy test magnesium capsules to determine whether you need additional magnesium, which you can take just prior to going to bed.

Now is also a good time to introduce you to the energy medicine question-and-answer area of my Web site (access from www.innersource.net and put "insomnia" in the search engine). One or two of the techniques described there is likely to be a good resource for you. Several of the techniques that help with insomnia are general-purpose exercises, and you have already been introduced to them, such as Expelling the Venom (page 151), Crown Pull (page 58), Triple Warmer Smoothie (page 107), Connecting Heaven and Earth (page 44), and Hook-up (page 62). Because hormonal imbalances may be involved in the insomnia that is often reported with menopause, doing the Daily Energy Routine with the Menopause Module shortly before going to bed may also help you get a good night's sleep.

Going through menopause in our culture can be a soul-shaking experience. But it is also a mythic journey into a deeper engagement with life, designed to lead to the wisdom of the crone and the profound ways she is called upon to influence her extended family and community. Knowing how to evoke your energies to support you in this journey is like going on a safari with hiking boots and proper gear instead of sandals and a parasol.

Chapter 7

Weight Management Tips

Fifty percent of [women in the U.S.] diet two or more
times each year with little or no long-term success
at reaching or maintaining their desired weight.

—WEIGHT WATCHERS

Every day the media confront us with an embarrassment of anorexic-looking heroines. They coagulate in our psyches as an unworkable yet coercive ideal that is hurtful to our health, our sense of well-being, and our pride in ourselves. Before using energy medicine to address my own weight challenges, my life was permeated with the familiar self-criticism, self-denial, weight loss/weight gain/weight loss/weight gain, and bottled-up hysteria that is entwined in this cultural insanity. Like many women, I was tyrannized by the ideal of slenderness.

But beyond those who feel fat at a healthy weight, it is also true that more than 150 million Americans weigh substantially more than what is optimal for their health, with at least 60 million of them being clinically obese. And these gloomy numbers have been skyrocketing in recent decades. How do we deal with this epidemic of obesity without swinging back to anorexic ideals that carry their own evils?

In addition to the blow to our vanity, excess weight leads to numerous health problems that could be avoided, from diabetes to heart attacks to strokes to liver failure. The vast majority of the 150 million–plus overweight individuals have tried many strategies to lose those pounds. But they run into a number of paradoxes. For instance, your body takes comfort in excess fat. It wants to survive the next famine, and extra fat is like cash in its emergency bank account. In fact, if you allow yourself to become extremely hungry, you throw off your blood sugar and metabolism, triggering the ancient response of storing whatever you eat as fat. Even sleep enters the picture. Your body tends to store more weight if you are sleep-deprived. So giving yourself permission to get enough sleep isn't just something you do so you can be more rested and alert. Think of it as part of dieting!

Hundreds of diet books come and go! Clearly, dieting is not the whole answer for most of us. In fact, dieting often has a paradoxical effect, turning on mechanisms that cause your body to produce fat cells and preserve the fat it already has. Dieting signals to your body that the famine that was universally dreaded by our ancestors has begun. Turn every morsel of nutrition into fat storage! Now!!!

Shelves of different diet books also reflect the fact that each of us is biochemically unique. The same food combinations that cause one woman to lose weight are stored on another's hips. While the "calories-in minus calories-burned equals calories-stored" theory is true for some people, it totally misses the complex genetic and biochemical realities of weight storage for many others. Thin people often assume that excess eating is the only cause of excess weight. End of story. The rest is lies and excuses. Their more portly companions, noticing that their own food intake is far less than that of their judgmental beanpole friends, are not convinced that this is the scientifically accurate conclusion. The fact is that a large proportion of people who are heavy eat less and exercise more than their thinner friends. Yes, totally unfair! No weight management plan fits more than a minority of people. And even those programs that attempt to account for individual differences provide general guidelines that often are simply inaccurate. You are too unique, and thus such generalities are not reliable.

In my experience, energy medicine can be extremely effective in weight control. Several energy systems are involved in metabolism, in the creation of muscle cells, and in the storage or burning of fat cells. Because these energy systems tell your chemistry what to do, you can manage your weight by managing these energies. In the sense that it is easier to control the temperature of your refrigerator by adjusting the temperature knob than by dumping in ice cubes, working with the *energies* that control weight is like having your hand on the thermostat.

I once led an ongoing weight and energy medicine workshop for 16 people. The group included six women who were referred to me by local physicians because their weight was posing a serious hazard to their health. Each was more than 300 pounds. Because I was still experimenting with using energy medicine for weight management, I offered to do the group for free under one condition. And that was that they not diet (I'm known locally as a very tough negotiator). We worked together for about 16 months, two sessions per week. Each of the six women referred by their doctors lost more than 100 pounds. And they kept the weight off.

I wish it were easy to put into a book the methods that work for weight management. I have found, instead, that there are few requests that come through my door that require a more individualized program or more careful monitoring than controlling weight. Successful treatment combines an understanding of the woman's unique biochemistry, energies, self-image, internal conflicts, and weight history. It is easy to underestimate the physiological pressure to keep fat on the body. Willpower may have little impact if your strategies do not take into account the energetic dimension of weight control. Plus, the approach needs to be continually monitored, as the strategies that work one week may need to be revised the next. Your body's energies are always shifting. Perhaps one day I will attempt to put all of that into a book, but it will not, alas, be a simple book. What I can offer in this chapter, however, are some of the techniques that seem to be effective for a reasonable number of people.

Rather than laying out a systematic program that pretends to address the unique biochemistry of each reader, I simply present them as "Weight Management Tips" in three categories: Tips for making healthy choices when your body is screaming "FEED ME," tips for reprogramming your mind to support healthy eating habits, and tips for reprogramming your body to maintain a healthy weight. While starting here with full disclosure that effective weight management using energy medicine often requires a highly sophisticated individualized approach, you may experiment with these tips and find that some are genuinely useful for you. For instance, if the energies moving through your spleen meridian are low, it is very hard to metabolize food properly and therefore to burn calories efficiently. So several of the "tips" offer ways of promoting a healthy energy flow in the spleen meridian. Simply tapping your spleen points (Figure 2-5, page 54) for about ten seconds before and after eating improves your metabolism. How cool is that! Consider each tip an experiment and discover the ones that are helpful to you. I hope you will find them useful and empowering.

WHEN YOUR BODY IS SCREAMING, "FEED ME"

The desire to place tasty food in your mouth is nearly as primal a biological urge as breathing itself. Your body is designed, in one of nature's striking examples of the body-mind connection, to cause you to crave the foods it needs when it needs them. Not to be outdone by nature, however, the food industry attempts to manipulate your body's natural preferences with artificial sweeteners and other miracles of modern chemistry, fooling your taste buds into responding to substances that are not good for you as if they were good for you. In addition, because of imbalances in the body or the psyche—often caused by stress, habit, or biochemistry—the signals sent to your brain regarding the need to eat do not match your body's actual need for sustenance. The result may be compulsive or anxious eating that not only forces you to have to process material you do not need, but also taxes your body and hurts your health.

Each of the Weight Management Tips addresses such dilemmas with energy interventions, from restoring physiological and psychological balances to what to do at the moment of temptation. The tips in this first set present straightforward steps you can take when your body is screaming, "FEED ME." If uncontrolled eating is an issue for you, experiment with these and keep the ones that work best in your back pocket. The techniques presented here address:

1. Stressed Eating
2. Comfort Eating
3. Anxious Eating
4. Compulsive Eating
5. That Hollow Hungry Feeling
6. Specific Cravings in the Moment
7. Changing a Pattern of Cravings
8. Overcoming Hunger
9. Subclinical Allergies

Tip 1: Stressed Eating

Don't eat when you are stressed! First unstress! Then enjoy your meal. When your body is in fight-or-flight mode, the resources are not available for proper digestion, making it hard to metabolize food or acquire the nutrients it has for you. The Five-Minute Daily Energy Routine (page 51) is a good way to bring yourself out of stress, at least enough to enjoy a meal. A quicker and simpler way to bring your body out of stress is the Neurovascular Hold (page 151). Another is the following (time—less than a minute):

Figure 7-1
Eat in Peace

1. Raise your hands high above your head, make fists, and on a slow, controlled exhalation, bring your arms down slowly and deliberately, fists in front of your body (see Figure 2-16, page 75). Open your hands when they reach the bottom.
2. Place your hands back-to-back and slowly and deliberately slide both hands up the center of your body (see Figure 7-1).
3. Lift your hands in front of your face and above your head on another deep in-breath.
4. Turn your palms outward, and circle them down to the sides of your legs with another out-breath.

Taking your arms down the central meridian releases tension buildup and calms the body. The upward motion of your hands "zips" this calmness into your central meridian. Lifting your arms above your head and then taking them down your sides is calming to your nervous system.

Tip 2: Comfort Eating

Sometimes we eat simply because we want to be comforted. Hunger is not even a factor. We may take in food because we are feeling lonely, depressed, lost, or overwhelmed. An alternative way to comfort yourself is to calm the energies that are involved in these emotions, all of which ultimately trace to triple warmer. A simple yet powerful way to calm these emotions is the Triple Warmer Smoothie (page 107).

TIP 3: ANXIOUS EATING

Other times we eat to quell feelings of anxiety. Again, hunger is not a factor. And again, guess which meridian is directing the energies: triple warmer! The Triple Warmer Smoothie can help manage the urge for anxious eating as well as for comfort eating. While other methods for dealing with causes of the anxiety may be called for, the following technique is another way of calming triple warmer, turning off anxiety's alarm response. This method also sends a signal up the central meridian, calming the central nervous system and providing immediate relief without the calories:

1. Place your middle finger on a point about two inches below your belly button.
2. Take a deep in-breath, and as you release the breath, pull your finger up to your navel, slowly and with pressure.
3. Repeat two more times.
4. Flatten one hand on your womb chakra, below your navel, and the other on your heart chakra, over the middle of your chest. Rest them there for several deep breaths.

TIP 4: COMPULSIVE EATING

Here are three simple techniques that can be used, individually or in combination, to interrupt the compulsion to eat what you don't need:

1. Place either thumb in your mouth and suck it while bending your index finger over the area between your lip and your nose (as a baby does) and rub this area up and down slowly with your knuckle. *This hooks up the central and governing meridians to create a force field that is comforting and also provides a minicranial adjustment that moves oxygen and cerebrospinal fluid through the head, shifting the focus from food to the comforting feelings that are elicited.*
2. Place one hand on the middle of your chest. With your other hand, tap the point that is between the fourth and fifth fingers, just above the knuckles,

toward the wrist (see Figure 3-3, page 108). Breathe deeply as you tap and think about the food you are obsessing about. *This is an acupressure power point on the triple warmer meridian that turns off the ancient, primal response to grab any available food out of a fear of scarcity.*

3. Read the instructions for Expelling the Venom (page 151). Instead of releasing anger or frustration on the outbreath, release the panic or urgency to eat. On the Zip-up at the end, use an affirmation such as "I am feeling peaceful inside" or "I am satisfied and pleased with myself."

TIP 5: THAT HOLLOW HUNGRY FEELING

The Sunrise/Sunset technique (page 221) not only stabilizes blood pressure, it also, like a sedative, impacts your chemistry in a way that produces an inner peace. As you lift your arms to the heavens, you also open to larger forces, allowing beneficial surrounding energies to come in and fill you. Another quick technique when you are tempted to use food to fill that hollow hungry space inside is called Maddie's Belly Scoop (time—about 20 seconds):

1. Place your hands at your waist with your thumbs on your back and your fingers on your front.
2. Take a deep breath in, and then pull your thumbs forward with pressure (Figure 7-2a).
3. As you release your breath, let your thumbs continue to slide forward toward your navel.
4. When they are about two inches from your navel, let your hands and arms swoosh forward with power, scooping out the energy and sending it outward (see Figure 7-2b). Repeat several times.

A third technique for countering the hollow hungry feeling is rooted in a 6,000-year-old Taoist practice that takes your attention to the part of you that feels empty.

1. Lie on your back and place the palm of either hand on your navel. Move your hand slowly in small clockwise circles, at first only brushing the skin. Slowly make larger and larger circles. Feel the warmth. As you continue, allow your mind to experience all the sensations that are produced in your

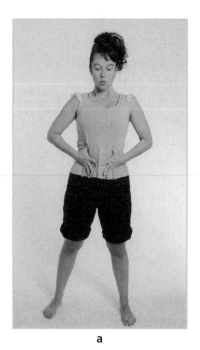

a   b

Figure 7-2
MADDIE'S BELLY SCOOP

body. Once the circles are surrounding your entire stomach area, begin moving inward again, toward the center, continuing in a clockwise direction.

2. Repeat the above sequence, this time using more pressure, and have your mind's eye watch the energies as they move.

3. Do the sequence one last time, now pushing in quite deeply, literally massaging your intestines. As you follow the sensations, imagine that excess fat is dissolving with each circular motion. *The circular motion quiets hunger, removes the "hollow" feeling, and stimulates peristaltic activity. Other benefits include tightening the flesh in the areas being massaged, strengthening the blood vessels, and helping the digestive system work more efficiently.*

TIP 6: SPECIFIC CRAVINGS IN THE MOMENT

A quick diversion when you are craving a snack your body does not need is to place one middle finger in your navel, push in, and pull up several inches with pressure.

Then pull down, back to the navel. Repeat, this time pulling out to the left side and back to the navel. Continue around until you have made a five-pointed star. *The technique frees blocked energies in your abdomen, providing its own kind of satisfaction, and breaking the focus that was going toward the craved delicacy.*

Another way to quell a craving is to do the variation of Expelling the Venom described in Tip 4 (step 3, page 261), but for the affirmation, say something along the lines of: "I feel *so* content and satisfied. Losing weight feels better than eating that pie."

Caution: *Watch the labels on what you ingest. For instance, some foods are manufactured with addictive sweeteners, such as leptin and ghrelin, that manipulate hormones to tell your brain you are hungry no matter how much you have eaten.*

TIP 7: CHANGING A PATTERN OF CRAVINGS

Many forms of hypnosis and self-suggestion have been applied for helping people manage their eating habits. Two of the best that I know are the Zip-up, with an affirmation (page 75), and the Temporal Tap (page 75).

If potato chips are your downfall, an affirmation you might use with the Zip-up might be "I feel empowered saying no to potato chips." With the Temporal Tap, you might state as you tap on the left side (negative phrasing), "I don't even like that processed taste of potato chips anymore," and on the right side (positive phrasing) tapping in, "I love being freed from the tyranny of the potato chip."

Another way to break an addictive craving is to completely change your body chemistry and then to use more psychologically oriented interventions, such as the Zip-up or Temporal Tap. For instance, doing a juice fast for a couple of days can have many benefits, including interrupting physiological habits so new ones can be put into their place. Get instruction from a knowledgeable health care provider or good health food store so you will do the fast in a manner that gives your body the least disruption and greatest benefit.

TIP 8: OVERCOMING HUNGER

In the best of all possible worlds, we would eat when we are hungry, stop when we are wonderfully satisfied, and maintain the ideal weight for feeling great and staying healthy. That is how we are designed. Sometimes in today's world, however, you may want to refrain from eating even though your body is legitimately hungry. Here are three simple techniques to make this easier:

1. Directly in front of the opening to your ear is a flap. Grasp it with your thumb and index finger and begin to twist and massage it. *Acupressure points that have been used for thousands of years to turn off hunger when food was not available are situated on that flap.*

2. I find what I call "hypothalamus imagery" to be a great focus when I am dealing with hunger. The hypothalamus is a gland that, among other things, regulates metabolism and the desire for food intake. I imagine a little guy in the back of my head who is at the controls of my hypothalamus and whose job is to take care of me by inducing me to want food to stave off the famine and store all the fat I could possibly need. I just talk to him. I reassure him, telling him I am safe, putting him at ease in every way I can. When he is at ease, I am at ease, my body is at ease, and I am in control of how I respond to my hungers and cravings.

3. Do the Abdominal Stretch (page 134), Connecting Heaven and Earth (page 44), Diaphragm Breath (page 220), or any other exercise that stretches your abdomen. *Stretching the abdomen opens space for energy to flow, providing energetic nourishment.*

Tip 9: Subclinical Allergies

Subclinical allergies are responses to certain foods that do not easily metabolize for you. The symptoms are much subtler than a skin condition or indigestion. These foods can cause weight gain that is disproportionate to their calories. Identifying such foods is yet another reason to become adept in energy testing so you can "test before you ingest." If you know your body has difficulty with a specific food and want to experiment to see if you can help your body adapt to it, tapping on your spleen points (page 55) as you chew it can help your body metabolize it more easily.

REPROGRAMMING YOUR MIND

At more than 300 pounds and just starting to work with me, Joan sent this letter:

Well, I know I should feel ashamed of myself, and sometimes I do. Sometimes I feel rotten about other people's assumptions about me. That I'm sloppy and dumb and I eat lots of starch and have no self-respect and I just don't care. But I have my moments. After I get home and I undress, I look in the mirror. I see rings of flesh, massive planes, mountainous plateaus. Suddenly, I'm in a glory of fleshy existence. I see a power in it, in my own body. Fertility. Abundance. A large unrestricted sense of life. I'm actually proud of the way I look. How dare people presume I'd rather be thin!

It is a great paradox that accepting your body just as it is can be the start of a wonderful transformation. Eighteen months after writing the above, Joan was at 125 pounds following a twice-weekly weight support group that focused on energy medicine methods and doing daily homework, but no dieting for the purpose of losing weight. Her doctor expressed amazement that the energy work reversed a serious metabolic disorder that other medical interventions had not been able to correct. The last time I saw Joan, 16 years later, she was still at 125 pounds. Her set point (discussed later) had gone south for good.

This achievement, interestingly, was at first hard on Joan's emotions. After attaining such a wonderful relationship with her body in its fullness, she became depressed as the energy work changed her metabolism and she began shedding weight faster than anyone else in the group. She lost respect for the people who liked her so much more as a slim woman than as a round one. She left her marriage after a similar dynamic produced disdain for her husband. Going from large to small was not an easy journey. Had I been more experienced, I would have addressed her self-image and her acceptance of the transformation.

Your mind is a powerful player not only in how you feel about your weight, but also in how much weight your body stores and in the behaviors that affect your weight. The following seven tips present energy techniques that can help establish a better partnership with your body in relation to weight.

TIP 10: I LOVE ME JUST THE WAY I AM

As the great American psychologist Carl Rogers used to say, "The curious paradox is that when I accept myself just as I am, then I can change." Loving and accepting your body is the best way to get it to cooperate with any intentions you might have for changing it. Trust me. In my weight classes, each woman had the homework assignment of looking at herself in the mirror and expressing appreciations toward each part of her body. It began, however, with focusing on a body part about which she had judgments or bad feelings. While looking and thinking about this body part, she went through the following instructions:

1. Face a mirror, place your thumbs on your temples, in the indentation just outside your eyes, and rest the other fingers on your forehead. Take three or four deep breaths while holding this position. *The points you are holding reduce stress levels. Holding them helps blow the negative charge you have been harboring about this part of your body.*

2. Take a moment to recognize this part of your body as a miracle of nature, a sacred achievement of evolution. Ask your body for forgiveness about your judgments, something along the lines of: "Forgive me for having been so mad at you and so judgmental. You have loved me unconditionally, doing the very best you could to serve me."

3. Now place one hand over the middle of your chest (heart chakra) and the other just beneath your navel (womb chakra), mindfully connect with the part of your body you are looking at in the mirror, and state a sincere appreciation for it.

You may have to reflect a bit to identify the kinds of gratitude that emerge when you recognize what this part of your body is doing for you, but the process will engage you with your body, and its parts, in new ways. Here are some appreciations I have stated:

"Thank you, arms, for reaching for what I want, for expressing my feelings, for your strength, for your ability to embrace."

"Thank you, legs, for walking me to wherever I want to go."

"Thank you, tummy, for digesting my food and for keeping me warm and safe."

"Thank you, breasts, for making milk for my girls when they were little, for making me feel sexy, and for bringing me pleasure."

Whisper loving-kindness to your body as you would to any other intimate.

Tip 11: Maintaining an Upbeat Feeling While Dieting

The concept of dieting is connected with the concept of deprivation. It is more accurate, however, to recognize that smart dieting is a way of *giving* your body what it truly needs. You are not fighting it, you are loving it. Any time you begin to feel sorry for yourself or cheated or deprived of a food you might crave in the moment, this exercise, the Three-Heart Hook-up, creates a sense that all is well.

You will be tracing three hearts: one over your face, one over your torso, and one in the larger field that encompasses your entire body:

1. Cradle your face in your hands, with the bottom of your palms touching at your chin and your fingers reaching up over your cheeks. Draw your middle fingers over to the center, between your eyebrows (the "third eye"), push up, and with a deep breath, trace a heart with your middle fingers, ending at your chin. Trace this heart over your face three times, breathing deeply.

2. Then draw your fingers from your chin down to your chest. Trace a large heart up and around your rib cage, ending by flattening and moving your

hands in front of your hips until your fingers touch. Then move your fingers straight up to the center of your chest and create two more of these hearts, again breathing deeply.

3. At the bottom of that heart, place your hands so they are back-to-back, thumbs against your body, and bring your thumbs up the center of your body and over the center of your face, separating as they move over your head, and creating as large a heart as your arms can comfortably reach in the space that surrounds you, ending with your hands touching again at the bottom of your torso. Again place your hands back-to-back and bring them up the center of your body, breathing deeply while creating two more hearts.

TIP 12: CHANGING A PRESENT TRUTH TO THE TRUTH THAT YOU WANT

The Temporal Tap (page 75) is a powerful method for changing self-limiting mental programming. You often have to do it every day for about a month before it locks in, but for a long-standing, deep belief, you should expect that some persistence might be required. In relation to weight, these may be very specific and in-the-moment beliefs, such as "I must have chocolate right now," or quite far-reaching ones, such as "all the women in my family are fat by the time they reach 50, and fat is my fate as well." The wording used with the Temporal Tap is very important, and it is also important that you find wording that fits for you. The following are simply possible wordings for the two examples just mentioned:

Freeing yourself from chocolate:

Left (negative phrasing): I don't even like chocolate anymore.
Right (positive phrasing): I love feeling free from that yearning for chocolate.

Changing your expectations about the future:

Left (negative phrasing): It isn't my fate to get fat as I grow older.
Right (positive phrasing): I am going to be slim and fit at 50.

Tip 13: Countering the Fear of Fat

The culture's messages about being overweight create a tyranny of fear that literally permeates into your cells. That tyranny can keep you from seeing yourself accurately, while creating a panic to diet in unhealthy ways. If and when it grips you, the following, in any combination, can help quell it.

1. While tuning in to the feeling, place either hand over the center of your chest. With your other hand, tap the hand on your chest in the valley between the ring finger and little finger, above the knuckle (toward the wrist). Tap on each hand for several deep breaths.
2. While tuning in to the feeling, hold your neurovascular points (page 105).
3. While tuning in to the feeling, do the Blow-out/Zip-up/Hook-in (page 74).
4. While tuning in to the feeling, do the Meridian Energy Tap (page 80)

Tip 14: Countering the Fear
of the Bathroom Scale

More women have told me of their terror about getting on the bathroom scale than you might imagine. Their scales have become triggers for self-doubt, emblems of emotional discouragement. For some women, this has become so extreme that they haven't weighed themselves for years. I've even been told, "That's why I don't get regular medical checkups—I don't want to see how much I weigh, and I certainly don't want anyone else to see it and record it." If thinking about weighing yourself sends you into fear or despair, then your relationship to the bathroom scale deserves attention. The following sequence can help you break the tyranny of that seemingly innocuous piece of equipment.

1. Stand in front of the scale. Do a Blow-out (see page 74). Blow out feelings of fear or despair.
2. Step onto the scale. Take a good look at your weight. Continuing to look at the numbers, breathe in deeply through your nose.
3. With your breath still held, touch the thumb of each hand with the index finger of that hand and press them together hard. Hold for several seconds. Let your breath out through your mouth very slowly as your fingers release.

4. Whatever the news from the scale, step off it and take another deep breath.

5. Place your left hand over the center of your chest and, with your right hand, tap between the ring finger and the little finger of your left hand, just above the knuckles, toward the wrist (see Figure 3-3). Tap for several deep breaths and then repeat on the other hand.

6. Do the Zip-up (page 75) with an affirmation such as, "I am grateful I have this gauge to give me good and helpful information."

TIP 15: COUNTERING DEPRESSION ABOUT WEIGHT GAIN

Sometimes it is very hard to be unattached about your weight. We may have done a great deal to free ourselves from the culture's tyranny about staying absurdly thin, yet when we step on that scale and find that it has been climbing upward, we may feel discouraged and depressed. The methods described for menopausal depression (pages 245–248), including stretching techniques, the Homolateral Crossover, the Mellow Mudra, and the Heart-Womb Connection, in any combination, can be helpful for countering any form of depression.

TIP 16: APPRECIATING PLATEAUS

However effective your efforts, you are going to hit plateaus—periods where progress seems to stop or even go backward. While these can be discouraging, plateaus are inevitable aspects of the ebb and flow of change. They are necessary to help your body regroup. One way not to be thrown off by plateaus is to anticipate and reframe them. You can recognize a plateau as a good sign, a reason to celebrate, and a reason to rest. It is a time to drop into gratitude for the progress that has been made. You can't push here. Your body's intelligence knows how long you need to stay at an interim spot before it can embrace further change. By accepting the plateau, you don't stress your body. Stressing your body triggers the triple warmer and can reverse progress. When you do reach a plateau, one energy technique that is always helpful is to strengthen the spleen meridian (see page 132). The spleen meridian governs metabolism, and keeping this meridian strong can, along with time, help your body get used to the weight loss rather than bouncing back to the earlier weight. Strong spleen energies may also shorten the amount of time you need to be in the plateau.

REPROGRAMMING YOUR BODY

While the techniques for reprogramming your mind may be vital for weight management, making shifts in physiological processes that impact weight, without the use of drugs, is one of the great strengths of an energy approach. The areas of focus in this section include:

17. Lowering Your Set Point
18. Improving Your Metabolism
19. Assimilating Food More Effectively
20. Activating a Sluggish Digestive Tract
21. Slimming Your Waist and Belly
22. Oxygenating Your Body
23. Providing First Aid for Low-Blood-Sugar Attacks
24. Countering Hormone-Induced Weight Gain
25. Keeping Your Thyroid Healthy
26. Eliminating Toxins

TIP 17: LOWERING YOUR SET POINT

The set point is a set weight the body attempts to maintain. Like a thermostat turning on and off the heat or AC to maintain a predetermined temperature in a room, the body tends to keep bouncing back to a given weight. If you manage to get your weight to be less than your set point, a host of physiological mechanisms kicks in to store fat, such as changes in metabolism that reduce the rate at which calories are burned. It doesn't matter if your set point is higher than you wish it to be. When your weight goes below your set point, you can feel starved, with images of food flooding your mind. Your body chemistry might compel you to eat a whole bag of cookies, and even then you might not feel satiated. This is nature in action, trying to prepare you for the next famine with absolutely no regard about your personal desire for that anorexic look.

Many weight-loss approaches recognize that lowering the set point would be an excellent strategy for managing weight. How to do this, however, is a matter of debate. While aerobic exercise, for instance, is known to have a positive effect, it isn't exactly a magic bullet. The set point is *set* by many factors. Some, such as heredity, body structure, and age, are fixed—you cannot control them. Others, such as health and stress level, fluctuate. What is not usually discussed is how decisively the set point is controlled by your body's energies. Triple warmer, the energy system that maintains survival habits, manages the set point. If chronic stress pushes triple warmer into a condition of perpetual threat, for instance, raising the body's weight may be a misguided protective response. Reducing chronic stress can cause triple warmer to lower the set point. The techniques for balancing your overall energies presented in this book can have the happy side effect of lowering your set point. When I learned how to strengthen my spleen meridian, which was always my Achilles heel, I lost 17 pounds in seven weeks without changing my diet. This wasn't even my goal, yet it happened. Maintaining a good balance between your spleen and triple warmer meridians, as emphasized throughout this book, is among the most important steps I know for keeping your set point stable and sometimes lowering it as well. You can also reset your set point more directly by using powerful self-suggestion techniques bolstered by stimulating energy points.

The Temporal Tap (page 75) is a powerful approach. You need to find wording that fits you and your situation. Take into account internal obstacles to losing weight. If the comfort brought by eating has become a substitute for other rewards in life, you must address this directly. Remember, use a negative wording when tapping on the left side and a positive wording when tapping on the right side. So while tapping on the left side, you might say, "I no longer eat to curb my sense of emptiness, and my set point does not resist going down." When tapping on the right side, you might say, "I am savoring every bite and my set point is dropping to [five pounds less than your current weight]." If having extra weight is at some level a way of protecting yourself against unwanted sexual advances, your wording on the left side might be, "I do not need to keep extra fat to set clear boundaries." On the right side, you might say, "I know how to assert a clear 'No!' as my set point drops to [five pounds less than your current weight]." Thousands of variations are possible, but 30 days of tapping with wording that fits for you can make a real difference. I've seen this happen over and over. Being able to lower your set point through self-suggestion and energy work is highly empowering.

Tip 18: Improving Your Metabolism

Another fast track to reducing weight is to get your body to break down food and burn calories more rapidly and effectively. Oxygen is a key to healthy metabolism (see page 249), and the Metabolic Breath (page 250) is a one-minute technique for helping your body metabolize food more efficiently. If your energy is also dragging, which often goes along with slow metabolism, chances are that you are in a homolateral pattern and will benefit from the Homolateral Crossover (page 67). Then do the following two or three times each day:

1. Do the Five-Minute Daily Energy Routine (page 51) and follow it with Connecting Heaven and Earth (page 44).
2. Place your thumbs on the triple warmer neurovascular points at your temples and the pads of your fingers on your forehead, on the bony area above your eyebrows (see Figure 3-1, page 106). Hold here softly for up to three minutes, breathing deeply in through your nose and out through your mouth. *Triple warmer's default setting is to keep fat on your body so you will survive if food gets scarce. It did not read your diet books. And the more stress you are facing, or the greater your sleep deprivation, the more fat it wants to store. Calming triple warmer is a way of telling it that it can relax its extreme survival strategies, including extra fat storage.*
3. Do the Spleen Meridian Flush and Tap (page 136) to speed your metabolism. *The spleen meridian governs metabolism.*

While these energy techniques address metabolism directly and effectively, there is also no substitute for regular aerobic exercise and vigorous stretching. Build them into your daily routines.

Tip 19: Assimilating Food More Effectively

Your small intestine's job is to extract the nutrients from food and assimilate them into your body. If the small intestine does not accomplish this effectively, you are required to take in greater quantities of food to get the same nutritional value. Your body literally "hungers" for the missing nutrients. It compels you to eat additional quantities because it was not able to absorb the nutrition that was available in the

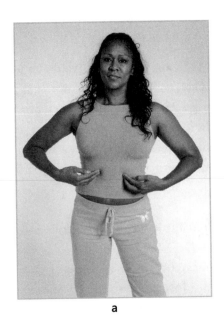

a   b

Figure 7-3
ASSIMILATION MASSAGE

meal you just had. This adds unnecessary weight. In addition to taking the steps discussed above for optimizing your metabolism, you can assist your small intestine in assimilating food more effectively by massaging its neurolymphatic reflex points two or three times each day. You can massage the two major sets of small intestine points in about 20 seconds:

1. Cup your fingers and massage along the bottom of your rib cage (see Figure 7-3a).
2. Grasp your thighs with your hands and massage your inner thighs with your thumbs (see Figure 7-3b).

TIP 20: ACTIVATING A SLUGGISH DIGESTIVE TRACT

If your body is not eliminating properly, it is often because the ileocecal valve—the valve that connects the small and large intestines—is not opening and closing as it should. When this occurs, and it is quite common, the digestive system's peristaltic rhythm is not fully supported, digestion becomes sluggish, and material that is meant to be eliminated can even back up into the small intestine. A series of folds in the far end of the large intestine, called the Houston's valve, may also become involved. While a variety of physical problems may lead to malfunctions in the ileocecal and Houston's valves, stress and tension are often the causes. If you are feeling bloated or experiencing difficulties in eliminating, massaging the ileocecal and Houston's valves as follows is likely to be beneficial. Resetting both valves creates a symmetry between them.

Massaging the Ileocecal and Houston's Valves (time—about 20 seconds)

Figure 7-4
VALVE MASSAGE

1. Place your right hand on your right hip bone with your little finger at its inside edge (your hand is over your ileocecal valve).
2. Place your left hand at the corresponding spot at the inside edge of your left hip bone (your hand is over the Houston's valve) [see Figure 7-4]. Exert pressure as you slowly drag the fingers of each hand up six to seven inches with a deep inhalation.
3. Shake the energy off your fingers with the out-breath and return to the original position. Repeat about four times.
4. End by dragging your thumbs from top to bottom, backward along the same paths, one time with pressure.
5. Close with the Three Thumps (page 53).

MASSAGING THE LARGE AND SMALL INTESTINE NEUROLYMPHATIC REFLEX POINTS

Another way to support your digestive system when it is sluggish is to massage the large intestine and small intestine neurolymphatic reflex points. Massage with

Figure 7-5
INTESTINE POINTS

pressure using circular movements. Proceed down the legs, staying two or three seconds at each point and then moving about half an inch farther toward the knees.

1. The small intestine points follow the inside seam, starting about a hand's width below the groin and ending about a hand's width above the knee (see Figure 7-3b, page 274).

2. The large intestine points extend from your hip to your knee and follow the outside seam of a pair of pants (see Figure 7-5).

TIP 21. SLIMMING YOUR WAIST AND BELLY

Anything that dissolves tension and allows energy to move through the middle of your body clears toxins and helps your metabolism. The following four techniques, in any combination, accomplish this.

1. Massaging the Large and Small Intestine Neurolymphatic Reflex Points (as described above).
2. Connecting Heaven and Earth (see page 44).
3. Abdominal Stretch (see page 134).
4. Sideways Stretch (see page 135).

Another very simple technique for slimming the waist is Tanya's Spiraling Swing and Slap (time—about a minute).

With your feet planted shoulder-width apart and your knees relaxed and slightly bent, twist to the right and allow your left arm to swing in front of your body, your right arm in back, both slapping your body with the twist (see Figure 7-6a). Swing back to center (Figure 7-6b) and to the other side with the same motion, swing, and slaps. If you have a tendency to get dizzy, looking straight ahead will prevent this.

a   b

Figure 7-6
TANYA'S SPIRALING SWING AND SLAP

TIP 22. OXYGENATING YOUR BODY

The diaphragm is like a bellows that opens and shuts as you breathe. With each inhalation, it creates suction that draws oxygen into the lungs. Stress can interfere with the diaphragm's operation, keeping the cells in your body from receiving optimal levels of oxygen. We actually have little conscious control over its actions, yet we can exercise and strengthen it with the Diaphragm Breath (page 220). This results in a better distribution of oxygen throughout the body. Among the many health benefits are better metabolism and reduced levels of fat storage.

TIP 23. PROVIDING FIRST AID FOR LOW-BLOOD-SUGAR ATTACKS

Any diet plan that subjects you to low blood sugar is not a good plan. When blood sugar drops, the body's energies become weak and disorganized. The pancreas governs blood sugar levels, and the pancreas is governed by the spleen meridian. If food is not immediately available, doing the Triple Warmer Smoothie (page 107), followed by tapping the spleen neurolymphatic reflex points (page 55), helps reorganize your energies. The Wayne Cook Posture (page 56) can then further stabilize them.

TIP 24. COUNTERING HORMONE-INDUCED WEIGHT GAIN

Hormones may be involved in a sudden weight increase or may be keeping weight on that you cannot shed. Medications may also disrupt your hormones in ways that affect your weight. Techniques from throughout this book, such as those in the Daily Energy Routine (page 51) or the Hormone Hook-up (page 220) or others in the Menopause Module (page 219), may help balance your hormones so food is metabolized more effectively. Such methods always bring some benefit to the body's energy systems. You can experiment to see which have the strongest impact. A quick way to stimulate the organs that process hormones is to firmly tap up the insides of your legs several times, from above your ankle to your groin, for about 30 seconds. This stimulates the liver, kidney, and spleen meridians. Then firmly tap down the outsides of your legs several times, again for about 30 seconds. This stimulates the stomach and gallbladder meridians. As you tap, breathe deeply in through your nose and out through your mouth.

TIP 25. KEEPING YOUR THYROID HEALTHY

Your thyroid impacts all your hormones, and it can become less effective as you age. It is often involved in the weight gain that may begin in the middle years. See the discussion on pages 231–233 for ways to keep your thyroid functioning at its best.

TIP 26. ELIMINATING TOXINS

The body is continually processing the air you breathe, the food you eat, and the stresses you encounter. Toxins are harmful by-products of these activities, and they may accumulate in your body. Toxins can hinder your health in numerous ways, including interfering with your metabolism. The techniques presented for moving toxins out of your body (pages 45–51) are generally valuable, and they can be of particular importance when holding excess weight is an issue.

FARE WELL

Just as energy methods can empower you to be master instead of victim of your own weight, I hope this book and the methods and perspective it offers serve you in becoming more the master of your own fate. An approach that attunes you to your body's energies does empower you. It allows for deeper levels of involvement—conscious and purposeful—with the master control centers of your body and mind. I have time and again witnessed how a shift in the energy systems that govern health resulted in renewed vitality, strength, and joie de vivre. These energies are a bridge between your physical body and your spirit, so health maintenance becomes a journey into the foundation of your being.

My intention with this book has been to lead you on a path of discovery and empowerment by guiding you through the energy systems that form the infrastructure of your physical body. On this journey:

- We began with an exploration of the ways that energy fields control every biological mechanism.
- You were introduced to the concepts that by shifting those energy fields, you can bring about desired changes in any physical condition on which you wish to focus.
- You learned a variety of techniques for stimulating your body's overall energy system to promote health, clear thinking, and well-being, packaged as a Daily Energy Routine, the energy medicine counterpart of a physical workout.
- You learned to apply energy methods to specific concerns, from managing stress and emotional distress to reprogramming deep-seated energetic patterns in body and mind, and then to issues that are of particular concern to women, including menstruation, sexuality, fertility, pregnancy, delivery, menopause, and weight management.
- In the process, you learned about the web of connections among your body's hormones, your energies, and your well-being.
- You learned techniques for managing those energies that can help your hormones function at their best, and I hope you have felt the benefits that good management of your energies and hormones can have on every aspect of your life.

All this was in the larger context of a culture that has been evolving and changing at a rate that by any measure is out of control. As the requirements of daily life pull us further and further from the natural rhythms of our body, energy medicine becomes an antidote, reminding us of who we are and reconnecting us with our deepest nature. And there, we find, lies the true gift of creation. We who give birth know in our hearts and bones and wombs how energy creates, acclimates, adjusts, and builds healthy bodies. As we step into a future where the feminine principle walks hand in hand with the physician, the statesman, and the educator, the new world we so deeply long for is being born. This book opened and now closes with an emphasis on the happy fact that every woman is an exquisite vibration, a spark of the divine archetypal feminine principle. As you embrace your deepest nature more fully, the energies at the foundation of your being shine forth into your body, your mind, and your world.

Appendix

Talking with Your Body's Energies[1]—
The Art of Energy Testing

The Chinese physician can detect imbalances in meridians
by feeling the pulses, but this is a sensitive touch, and it may
take ten to twenty years to develop proficiency with it.

—JOHN THIE
Touch for Health

nergy testing, on the other hand, can be learned quickly. You don't need any great intuitive powers or an ability to "see" energies. In fact, when I learned energy testing, what excited me the most was that it gave me a way to demonstrate to others what I was seeing in their energy field. With energy testing, you can reliably assess, in any given moment, the body's unique energies and energy fields and their ever-oscillating fluctuations.

Your body is, in fact, a cascading fountain of energy systems, remarkably complex, exquisitely coordinated, and entirely unique. That is why no book can tell you exactly what you must do to thrive. Not only does every person carry a distinguishing energy, so too does every cell, every organ, and every system of the body. Just as your thumbprint, your heart, and your brain are unique, so are the energies of each. These local energies speak their own language, and they also share a universal language. Energy testing is a precise tool for translating these languages, phrase by phrase. It is also the case, and all my students eventually experience this, that the more you work with your own energies and the energies of others, the more strongly you develop your ability to intuitively sense them.

THE ART AND SCIENCE OF ENERGY TESTING

As you develop your sensitivities to the subtle energies swirling within and around you, it will be extremely useful to have a tool that is tangible and palpable rather than to be left to rely totally on intuition. Enter energy testing. Energy testing, developed as "muscle testing" by George Goodheart, the founder of Applied Kinesiology, and meticulously refined by his protégé Alan Beardall, is a concrete and very tangible procedure. It allows you to determine whether an energy pathway is flowing or blocked, whether an organ is getting the energy it needs to function properly, or whether an outside energy (such as the energy of a particular food or a suspected toxin) is harmful to your system. I call the procedure "energy testing" instead of "muscle testing" because the test really assesses the *flow of energy* through the body's meridians rather than the *strength* of the muscles used in the test.

When I was first learning energy testing, a problem whose solution had eluded me all my adult life was solved. After years of experimentation, I had not been able to figure out why I couldn't manage my hypoglycemia or my weight, even though I was diligently taking the steps that should have controlled both. My spleen meridian consistently tested weak. The pancreas, which controls hypoglycemia and often weight, is on the spleen meridian. By working with the spleen meridian, I was able to counter my lifelong hypoglycemia. I was impressed. I also lost seventeen pounds without changing my diet. I was *really* impressed. Over the years, energy testing has been priceless within my personal as well as professional life.

While MRIs, EEGs, and CAT scans provide vital and often lifesaving information about the body and its energies, I have yet to find a medical device that can reliably make the *subtle* determinations afforded by energy testing. Built into each of us, however, is all the equipment we need to ascertain which energies are good for a person and which are not. Energy testing is always available, night or day, and requires no instrumentation. If you practice it regularly, it can begin to feel almost instinctive. It is a tool that lets you ask your body for information about what it needs. And it lets your body answer in a language with a small enough vocabulary that you can master it.

In fact, energy testing is quite easy to learn. Deceptively easy. As a result, many people apply and misapply it casually and often inaccurately. Misapplied, it may reveal more about the beliefs of the tester, the fears or hopes of the person being tested, or other factors that have nothing to do with the information being sought. Many people have encountered energy testing by an inexperienced practitioner, have been given flamboyant explanations of its capabilities, or have seen it used more like a parlor trick than a tool for self-knowledge. I want to lift energy testing in your mind to its rightful place, which is between science and art. It is an art to learn how to energy test reliably. Once you have, it becomes a dependable barometer of your body, your energies, and your environment.

If I had the power to impact the medical profession in just one way, it would be that physicians add energy testing to their diagnostic tool kit for determining the choice and dosage of medications. Iatrogenic illnesses—disorders induced by medical treatment—are among the most serious problems in medical care today. Energy testing could significantly reduce their incidence. And if I could choose just a few ways for this book to impact your lifestyle, one of them would be that you use energy testing to determine what foods, vitamins, and supplements you should, and should not, take into your body. In maintaining health, knowledge is power, and because each of us is unique, energy testing is vital in my work.

Energy testing allows you to assess the state of your own or another's energies, identify imbalances, and tailor the procedures presented throughout this book to suit your own unique needs or those of someone you care about. A woman with multiple chemical sensitivities and a history of life-threatening allergic reactions needed an antibiotic for a strep infection. When she went to the drugstore to get the prescription filled, she asked if she could bring it back if she proved to be allergic to it. The pharmacist of course said that would not be possible, but she was on a limited budget, and she was adamant that she not wind up buying medicine she couldn't use. She knew how to energy test, and she convinced one of the drugstore staff to energy test the medication on her. She tested strong and purchased the medication.

The pharmacist was so amused that the story got back to me. The story had an unexpected but instructive ending. The medication seemed to be effective, with her symptoms subsiding within a few days, but it was supposed to be taken several times per day for a full course of ten days. At about day six, the woman's neck began to itch, her stomach and ankles swelled, and her heart began to fibrillate. She recognized this as a severe allergic reaction and had her granddaughter energy test everything she had recently ingested. The antibiotic now tested weak. After she stopped taking it, the symptoms soon subsided. As often happens for people with hyperallergenic tendencies, she had built up an intolerance to the medication. Does this mean that the original energy test was inaccurate? No. The body is a dynamic system that is always in flux. Intolerances can and do develop. Because energy testing is quick, free, and always available, it is an extraordinarily useful tool for obtaining current information about your body's changing needs.

How Reliable Is Energy Testing? A 1984 article in the journal *Perceptual and Motor Skills* is one of the first published laboratory studies that supports the potential value of muscle testing, or energy testing.[2] The investigator, Dean Radin, later commented: "To my surprise I found that in double-blind tests, people were slightly weaker when holding an unmarked bottle of sugar than they were when holding an unmarked bottle of sand of the same weight."[3] A literature review in 2007 of more than 100 subsequent studies, including 12 randomized controlled trials, led the reviewers to conclude that there is scientific support for the method.[4] Some of the studies were quite impressive. An objective instrument called a "force transducer" has been developed. The force transducer has demonstrated a correlation between measured muscle resistance and the practitioner's assessment, and it has also been used to show that experienced practitioners using muscle testing had reliability and reproducibility when their outcomes were compared with one another's.[5] Another persuasive and well-controlled study compared muscle testing with measurements on computerized instruments and showed the difference in muscle firmness as the subject made congruent or noncongruent statements (truths or lies) to be highly significant (at the .001 level of confidence).[6] Some studies have yielded contradictory results. One showed significant agreement among three examiners when the test used the piriformis or the pectoralis muscles but not when the test used the tensor fascia lata muscle.[7] Other studies did not find the method to be reliable. Because there are so many ways that energy testing is conducted, and so many nuances within a single energy test, much more study is needed to determine how the method works and under what conditions.

While research confirmation always lags behind clinical experience, some nuances that must be incorporated into both clinical work and research studies have been established. For instance, different testers consistently reported the same results on the same clients when one set of instructions was given the client regarding resisting the tester's pressure, but not when another set of apparently valid instructions was given.[8] Other studies have investigated the physiology of the test. For instance, muscles that show weakness in an energy test register different voltage levels than muscles that are simply tired, so the test is measuring an internal shift that is different from fatigue.[9] Moreover, energy testing correlates with electrical activity in the central nervous system, so the information gathered during an energy test reflects brain activity, not just the state of the indicator muscle.[10]

Although more definitive research is still emerging, I have introduced energy testing to tens of thousands of individuals during the past three decades, and I have feedback from hundreds of them—who have returned to classes, written letters, and scheduled private sessions—that energy testing has proven a useful and accurate way of obtaining information about the body's needs. My experience has consistently been that the energy test corresponds with what I see in a person's energies. Information from the test also suggests where to work on the body, and the subsequent results have again and again confirmed for me the validity of the tests. The client frequently confirms the test as well, as when the test shows a weakness on the bladder meridian and the client immediately responds, "I'm just recovering from a bladder infection!" Some caveats apply, however. While this information is enormously useful, there are enough variables in an energy test that it should always be interpreted in the context of other sources of information. And unless a practitioner is highly experienced and proficient, decisions about medication should not be made on the basis of an energy test alone, and even then, only in consultation with the physician prescribing the medication.

A Biological Basis for Energy Testing. Your nervous system is a phenomenally sensitive thirty-seven-mile-long antenna, and your body's energy systems reverberate to the external energies that come into their range. Everything from the food you eat to the people you encounter carries its own frequency and affects you. While most of these vibrations exist below the threshold of your awareness, your body resonates to some and tenses against others. As a result, you will embrace the energies of some foods or some people while rejecting the energies of others.

Your sensitivity to these external energies is reflected in an energy test. The frequency of the substance being tested affects your nervous system, and this is reflected by the resistance in the muscle used in the energy test. Many kinds of energy can be measured scientifically, and I believe that energy testing can also discriminate among subtle energies that cannot be detected by existing scientific instruments.

Because energy testing detects the vibrational impact of a substance on your nervous system, subtle distinctions may be identified that could not, for instance, be revealed by a blood test. Foods that may appear identical, such as two apples— one organic, one not—may have very different vibrations and impact your energy system differently. I personally test strong for raw milk, weak for pasteurized milk, and very weak for skim milk. Natural foods are balanced within themselves, and when we remove a portion of that food, this balance may be corrupted, and the food's vibration altered. Your body then has to assimilate a vibration that has been skewed, which may upset its own balance or at least prevent it from gaining the full nutritional value of the food. It is very hard to be smarter than Mother Nature as we tamper with foods that sustain us, but energy testing can tell you if your body's vibration is in harmony with the vibration of a food or vitamin.

LEARNING TO ENERGY TEST

You can learn energy testing in a few minutes, but to master it, you need to incorporate it into your kinesthetic skill bank, like learning to ride a bicycle. Even as you begin to master the simple steps of energy testing, please understand that it is only with practice that you will be able to control for extraneous influences and attune yourself to the subtle distinctions that make an energy test accurate. The pressure you use in energy testing a child, for instance, is different from the pressure you use to energy test your sister, which is yet again different from working with a football player from your local college. But when your energy aligns properly with the energy of the person you are testing, the test will work regardless of the other person's strength. I became a bit of a legend in the Ashland High School athletic department when the coach brought me in to demonstrate energy medicine to the football team. These guys were not going to let this middle-aged blond lady budge their arm. But none of them could hold his arm firm after I fluttered the meridian that governed the muscle I was pushing against. Then they all wanted to know how to do this to the other team.

Being energy tested reinforces the link between your brain and subtle energies in your body, establishing new levels of internal communication. New areas of self-awareness begin to unfold. Many people find that they intuitively know what the result of an energy test will be before they apply pressure to the other person's arm. It's not like guessing before the test but rather launching a communication in which your awareness is working in tandem with subtle energies.

While energy testing is usually done with a partner, you can perform an energy test on yourself, and I describe that procedure as an alternative. For learning the process, however, it is better to have a partner. I cannot state too strongly how valuable it will be for you to push through any shyness or reluctance to involve another person in your learning here. Whether it is someone already close to you or just a casual acquaintance, it will be a gift for you both. We tend to touch people for affection, for sexual gratification, or in anger. Another very important reason to become comfortable with touching one another is for healing. It is an altogether different kind of touching. Beyond opening a door to new perceptions, feelings, and understanding, healing touch can save your life!

Energy Testing with a Partner. Every muscle, every meridian, and every organ in your body can be energy tested. A meridian is a fixed energy pathway that distributes energy to and from at least one organ. We will begin with a single test that you will be able to use in many contexts. This test determines the way the energy is flowing along the spleen meridian—the energy pathway that passes through the spleen and the pancreas. The spleen is involved with the immune system. It also determines whether the body will be able to metabolize a particular food, emotion, thought, energy, or other external influence. The spleen and the pancreas are both involved with food metabolism, blood sugar levels, and the mood swings associated with them. Both organs and their shared meridian influence your general energy level, and are extremely responsive to stress. For all these reasons, testing the spleen-pancreas energies can answer many questions you may have about the way your body will respond to something you are considering ingesting or otherwise bringing into your life. This is also an excellent gauge for assessing your body's overall health.

Because preconceptions can affect an energy test, you don't try to guess what the results will be. Subtle energies are responsive to your thoughts, so clear your mind as well as you can prior to conducting an energy test. If either of you is thirsty, begin with a drink of water. Water conducts electricity, and dehydration interferes with the flow of energy in your body. Avoid eye contact during the test, as this can make it more a test of interpersonal dynamics than what you are intending to test. Remove cell phones or other electronic devices, crystals, and heavy jewelry. Also be sure to ask if the person being tested has any injuries that might be aggravated by having pressure applied to the arm being tested, and if so, use the other arm. Having the person place his or her thumb and second finger on the points where the neck meets the head, about an inch from each side of center, stops the influence of thoughts and beliefs on the outcome. To energy test the spleen meridian:

1. Both of you take a deep breath. With your exhalation, release your expectations.
2. The person being tested places either arm straight down the side of the body with the thumb touching the side of the leg, fingers pointing down.

Figure Ap-1
SPLEEN ENERGY TEXT

3. The tester slips an open hand between the body and the arm, just above the wrist, and rests the other hand on the person's shoulder.
4. The tester then asks the other person to hold his or her arm firm, elbow straight. I often use words such as, "Hold your arm secure next to your body" or simply "Hold."
5. With an open hand and with pressure from one to two seconds, the tester slowly pulls on the arm (see Figure Ap-1).

With neither person straining, the arm will either pull away relatively easily or it will feel locked in place. When pressure is applied, a muscle with energy flowing through it may also move a fraction of an inch, but it will bounce right back. Do not struggle so much to hold your arm firm that you involve other muscles. If you are the tester, do not struggle to pull the other's arm away. This is not a competition, nor is it about muscle strength. If the energy is flowing freely, the arm will stay locked in place or give a bit but bounce right back when you pull on it.

If the Spleen Energy Test shows the meridian to be weak, strengthen it by vigorously tapping or deeply massaging the spleen points that are illustrated in Figure 2-5 (page 54) and then retest. In fact, if you find yourself overly susceptible to exhaustion, infection, or illness, one way to help keep your spleen meridian and your immune system strong is by routinely doing the Triple Warmer Smoothie (page 107) and then tapping your spleen points. Immediately after this sequence, the Spleen Energy Test will probably indicate that the meridian is strong. If the test still shows it to be weak, your energies are probably quite scrambled. Don't be alarmed. Do the Daily Energy Routine (plus the Homolateral Crossover) presented in Chapter 2 to unscramble your energy field, and then return here and continue. It is also important that the person conducting the test be relatively balanced, so some practitioners routinely do the Three Thumps (page 53), Cross Crawl (page 56), or Wayne Cook Posture (page 56) along with their client prior to performing an energy test.

If the muscle holds firm, you can challenge it, and this is particularly useful while you are learning the procedure. This is a way of determining the optimal amount of pressure to apply to distinguish whether the meridian's energy is flowing strongly. One way to calibrate this is to make a statement such as "I am wearing a blue shirt," where the truth of the statement is obvious to both parties. Immediately after the statement has been made, perform the energy test. If it is a true statement, the meridian will test strong, and you have a gauge of the pressure that can be applied when the energy is flowing. If it is a false statement, the meridian will not hold and you have a gauge of the pressure that indicates a weak meridian. Similarly, if you have the person think of something pleasant or joyful and immediately do the test, the meridian will generally stay firm. While holding an embarrassing or frightening thought, the meridian will generally lose its firmness.

If the meridian holds firm no matter what you do, or if it is chronically weak and will not be strengthened by tapping the spleen points or by the Daily Energy Routine, use the General Indicator Test described below. Sometimes, however, the body's energies are running in "irregular" patterns that need to be corrected before an energy test will be accurate. The energy moving through a muscle may be "frozen" or "submerged," or its polarity may be reversed. While more advanced than the scope of this book, instructions for working with such energies are available on the Web site of the Energy Medicine Institute (www.energymed.org).[11]

The General Indicator Test. The spleen meridian test you just learned is particularly useful for determining your response to foods, supplements, and environmental conditions. If, however, the spleen meridian is chronically weak, or if irregular energies are moving through it, it will not serve well in a general-purpose test. The General Indicator Test is a second energy test that may be used in virtually any situation because it does not isolate a specific meridian but rather indicates a more general disturbance in the energy field.

The basic principles for both tests are the same. The only difference is in how you hold your arm. To do the General Indicator Test:

Figure Ap-2
GENERAL INDICATOR TEST

1. Place either arm straight out in front of you, parallel with the ground, and then move it 45 degrees to the side. Your elbow should be straight and your hand open, palm facing the floor, or thumb pointing down, as in Figure Ap-2.
2. Have your partner place the fingers of an open hand on your arm, just above your wrist, resting the other hand on the other shoulder (see Figure Ap-2). As you hold firm, your partner pushes down slowly for up to two seconds and only hard enough to determine whether there is a "bounce."

As with the Spleen Meridian Test, you need the muscle to initially stay firm to get an accurate test, and you may strengthen the flow of energy through the muscle by using the Daily Energy Routine. You may then determine the amount of pressure to use by testing a true statement and a false statement. And if the muscle will not weaken or will not strengthen, more advanced methods may be needed.

Energy Testing Without a Partner. Because I sometimes wanted someone to energy test me when no one was available, I hit upon a solution while I was first experimenting with energy testing. I went to a sporting-goods store and found a dumbbell that I could lift if I held it straight out in front of me when I was holding a comforting thought but could not lift when I was holding a depressing thought (thoughts generate subtle energies that impact our bodies, and comforting as well as depressing thoughts affect the muscles).

I placed the dumbbell on a dresser that was the height of my shoulders. I would put my arm out in front of me, grab the weight, and try to lift it. If I wanted to energy test for a specific food or vitamin, I would hold the substance in one hand and try to lift the weight with the other. The energy of the food would affect my energies as decisively as an uplifting or depressing thought. I could find out whether the substance was having a positive or negative impact on my energies by my ability or inability to lift the weight. Because dumbbells exert a steady pressure downward, they can provide a reasonably objective measure of what is being tested. The critical physical factor for this self-test is finding a dumbbell with the correct weight—a weight you can lift off the desk or other surface while thinking a positive thought but not while thinking a negative thought. You can get the exact weight you need for this test by using a one-gallon water jug. Experiment with the amount of water needed to just be able to lift it off a surface with your arm straight out while keeping a comforting thought active but not a disturbing thought. Alternatively, hang your arm straight down while holding the weight and, with your elbow straight, lift it to the side, along the path you'd use if a partner was testing you.

Numerous other ways for energy testing without a partner have been devised. Because most of them require the person to simultaneously exert and resist pressure, to be both the tester and the tested, it is very difficult to assure an unbiased result unless the person is highly experienced. The most effective method I know has you use your body as a pendulum rather than exerting pressure on a muscle. While standing, place the item being tested at your stomach, holding it evenly with both hands, and bring your elbows in so the sides of your arms touch the sides of your body. Then bring your feet together, facing straight ahead. Become still, centered, take a deep breath, and release. If after a few moments you feel yourself pulled toward the item, that is, falling forward, it indicates that the item is in harmony with your energies (see Figure Ap-3). If you find yourself falling backward, away from the substance being tested, it indicates that the item is in conflict with your energies. It is as if, at a subtle level to which you must attune yourself, a substance that is in harmony with your energies has a magnetic attraction pulling you toward it while a substance that is in disharmony repels you. In this test, you need

Figure Ap-3
SELF-TEST

to be particularly careful about your wishes or pre-conceived ideas influencing the results, and the instructions for fine-tuning your energy-testing abilities with a partner, beginning on page 299, can be adapted for this test as well.

Practice Energy Testing at Your Next Meal. Although there is a different energy test for each of the fourteen meridians, the test for the spleen-pancreas meridian can serve you well in innumerable contexts. The test is based on the latissimus dorsi muscle, and it is particularly attuned to food metabolism. You can, however, use the Spleen Meridian Test as a *general indicator test* to find out about almost anything that is going on in your body.

Your next meal can become a personal workshop for practicing energy testing. Energy test each of the foods you are planning to eat. Touch the food and allow your other arm to be energy tested. If you lose your strength upon touching the food, your body's vibration is not in harmony with the food's vibration. This may mean the food is never good for your body, the food is not good for your body right now, or you have an allergy to the food.

Testing the food at various points in time will give you information about whether it is always, sometimes, or never good for you. Because everyone's chemistry is unique, nutrition is an entirely individual matter. One person's vitamin is another's poison. If the energy of a food, vitamin, or supplement doesn't match the energy of your body, you will not absorb and metabolize it even if all the experts in the world say you need it. Even good food is toxic if its vibration triggers your immune system to tie itself up in a defensive reaction. Such food allergies often go undetected, but not without accumulating damage. Energy testing can help you know what your body needs at a given moment, and it can help you develop a superb nutrition program for your unique body.

One caveat is that if your adrenals are exhausted, sugar and caffeine will often test strong. Even if a substance is generally harmful for you, if your body is needing the burst of energy it can supply, the test may be misleading.

Energy Testing Your Way to a Better Diet. For many people, learning to energy test for food changes their relationship to what they eat. You can test to see if your body is going to like a particular food. You can also get children involved with picking out foods you buy at the grocery store or energy test them before you cook a meal. It can be a game. They will enjoy it, at least until the first time they energy test weak on something they really want. But often, even if they are begging for something, seeing their arm lose its strength on an energy test of a junk food will bring laughter and a reduction in the tension.

Of course they may want to eat it anyway, but I have seen kids become less interested in foods that consistently cause them to lose their energy. The best results happen if they make the connection that losing strength on an energy test may mean that the food being tested may reduce their energy level and leave them feeling bad. But you have to be honest. A healthy body may be able to metabolize a certain amount of junk food, and it would undermine the process to overpower the child with your strength or your mind. Use the guidelines presented here to be sure the energy test is accurate, and particularly that it isn't being influenced by your beliefs or your child's desires.

Finally, let children draw their own conclusions about how eating food that energy tested weak makes them feel. When they can authentically relate energy testing to their life—that it isn't just a trick or a frivolous game—energy testing becomes a source of empowerment, a way to get useful information and biofeedback. They might also like turning the tables on Mom and Dad and energy testing you on that cheesecake or cigarette. And why not? You may be interested to find out for yourself whether that daily glass of red wine really is good for your heart. You can also use energy testing to assess the likely reaction of your own body or your children's bodies to vitamins and other food supplements.

How to Energy Test an Infant, a Pet, or Someone in a Coma. "Surrogate testing" allows you to test someone who is not able to offer resistance in an energy test. If someone is too sick to use his or her own strength or is mentally impaired or too young to follow your instructions, surrogate testing can provide valid information. This will even work with a pet.

It can also be used when you cannot get an accurate test on someone. For instance, because the energetic links with family members are so complex, you may find that you can get accurate results on everyone but your spouse or your child.

Or sometimes a person is muscular and macho, and recruits auxiliary muscles so that he or she will not look weak on the test. Surrogate testing bypasses these difficulties. Along with the person (or animal) being tested, it requires two others (the surrogate and the tester). If you are conducting the test:

Figure Ap-4
Surrogate Test

1. Have the two individuals hold hands or make other physical contact.
2. Energy test the arm of the "surrogate," the person touching the one you want information about. The results of the test on the surrogate tell you what is happening with the other (see Figure Ap-4).

This seems bizarre at first, but it is simply based on the way energy flows. If you want to know, for instance, if a particular food is causing an adverse reaction in an infant, you can be a surrogate for the infant. With the food on the infant's skin, place your hand over the infant's hand or abdomen and have someone energy test your other arm. If the test shows weak, the energy of the food is probably not compatible with the infant. If this seems too strange, please don't take my word for it. You can demonstrate its validity for yourself by first having yourself energy tested on a number of substances and then having someone surrogate test you on the same items.

What You Can't Energy Test. An energy test tells you whether the energy is flowing through or inhibited in the muscle being tested. Some people use energy tests to ask the body questions that go far beyond the scope of the procedure. "Is this condo a good investment?" "Will I enjoy going to Machu Picchu for my next vacation?" "Should I take the next Eden Energy Medicine Five-Day Basic Workshop?" I have seen more nonsense emerge from the misuse of energy testing than I like to think about, and it is an embarrassment to the field how many practitioners casually use energy testing to answer all sorts of bizarre questions.

Efforts to ask the body questions about the future seem particularly ludicrous to me: "Will I be cured of this illness by next month?" While I may have a lot of information to make an educated guess about such a question, an energy test is not a source I rely on for that information. I believe there is fate, there is free will, and there is circumstance. An energy test about the future assumes that it is all fated. But free will, unpredictable circumstances, and shifting relationships with other people all converge with whatever may be fated. This is why readings from even the most talented psychics are only a percentage game. Many factors influence the single question being asked.

But even questions about treatment are very tricky. "Should this problem be treated with the meridians? The chakras?" Energy tests can be influenced by many factors, even for people of the highest integrity who do everything they can to get their beliefs, hopes, or expectations out of the way. It is challenge enough not to exert an influence based on your beliefs, desires, or unconscious expectations. This is true for any energy test, but much more so when it is based on a verbal question rather than letting the body respond in its own language.

As I write this, I am having images of colleagues who ask questions before an intervention such as, "Is this for the body's highest good?" or "Do I have permission to proceed?" I ask these questions intuitively, but I do not energy test for them. However, practitioners I really admire do ask these questions and test for the answers so respectfully, almost as a prayer, that the questions themselves set an energy field of honor and respect. So the energy test itself serves a very powerful purpose of creating a shared energy field for healing.

So while emphasizing the pitfalls of asking verbal questions of the body, I do not want to dismiss the practices of colleagues who have found ways to make this work for them. Some practitioners have learned to use verbal questions followed by an energy test as a way of tuning in to a higher source of information, so it becomes for them a way of channeling in to their intuition, a special bridge to the truth of the situation. For people who have developed this to a refined art, it can be quite trustworthy. But for our purposes here, I simply want to present energy testing within the scope that I use it, which is much more traditional—checking, under specific conditions, the relative strength of an indicator muscle to determine the energetic state of a meridian or other energy system. While there are many truths and many ways to get at them, I believe that energy testing after asking the body a verbal question is one of the trickier of the bunch.

TESTING THE IMPACT OF THE PHYSICAL ENVIRONMENT ON YOUR ENERGY FIELD

The moment you come into contact with an external energy, before you are even dimly aware of its impact on you, your own energy field is already responding and adjusting to it. By simply bringing something near you and doing an energy test, you can discern its immediate impact on your subtle energies. I have many times arranged to meet a client at a supermarket for the purpose of energy testing to see which foods elicit a positive response and which do not. Even different brands of the same food or vitamin may affect you in different ways. It is also possible for our dietary needs to change from one stage of our lives to the next, but a tremendous amount of useful information can be garnered from a trip to the grocery store.

Your first energy tests are experiments for you and a partner, and you will both become more proficient in your ability to perform accurate energy tests as you proceed. For now, play with energy testing. Have your friend test you when you walk into your favorite part of the house and then when you walk into your least-favorite spot. Have a third person think "negative" thoughts while in your proximity. Find out how that affects your energy. Then have someone give you a genuine smile. Test again. Explore how the television affects you when you are two feet from it, then eight feet. Experiment with different setups of your computer or your office. Discover how your energies are affected when you are exposed to a particular piece of art or music.

As you explore your environment using energy testing, do not be concerned about temporarily bringing something that weakens you into your own energy field. Because we are continually entering fields of energy that affect us, the body—at least when it is relatively healthy—adjusts to the initial impact of a disturbing energy field by quickly rebalancing itself. In fact, one definition of health is how readily your body can adapt to a spectrum of environmental conditions. You can use energy testing to explore any element of your environment that stirs your curiosity. By having fun with it, you will be developing a strong aptitude with this invaluable tool.

FINE-TUNE YOUR ENERGY-TESTING ABILITIES

I have long sensed that energy testing does more than simply provide information. In an energy healing session, it feels like the early part of the treatment, as if performing the test somehow engages my energies and the client's energies within a healing context. Energy testing immediately directs healing forces to the area being tested. It is as if the energy test focuses your subtle energies in a way that prepares them for the work to follow. I have found that the identical procedure yields better results if I have energy tested it first. How can this be?

Because subtle energies are affected by thoughts and intentions, and because they are nonlocal, the energies of healer and client start mingling as soon as an energy test is even contemplated. We cannot be truly objective in any test. Not only does the act of observing influence the observed in quantum physics, subtle energies are ultrasensitive to many influences. Yet we can learn to control for our hopes, fears, and expectations, and create a container within which an energy test provides information that is very useful and quite precise. In fact, energy testing with a partner can reveal a deeper level of information than can any inanimate instrument. The complex interactions of tester and tested evoke two-way feedback that can make the energy test an incomparable dance of exploration for the purpose of obtaining trustworthy information. Having now trained hundreds of students whom I know to be capable of providing reliable energy tests, I can tell you with confidence that this skill can be yours, and that it is a skill worth developing. The following tips will increase your accuracy. Practice them now, or proceed with the earlier sections of this book and return here when the time is right.

1. **Maintain a Beginner's Mind.** Your mind affects your energy field, instantly and potently. To factor out preconceived ideas about what the test will reveal, approach energy testing as a contemplative practice, coming into a centered, meditative "beginner's mind." If you are concerned that your hopes or preconceived ideas, or your partner's, may be getting in the way of the test, the following procedure can energetically disengage them.
2. **Neurologically Disengage Expectations.** This approach for neurologically disconnecting preconceived ideas from an energy test, taught to me by Gordon Stokes, my first Touch for Health instructor, may seem implausible,

but over the years I have found that it works. It can be used by the tester, the tested, or both.

a. Take the thumb and middle finger of one hand and place them in the two indentations where the back of your neck meets your head.

b. With your other hand, proceed with the energy test (see Figure Ap-1, page 290).

These indentations are primary headache points in traditional Chinese medicine, but holding them does more than relieve headaches. The energy that travels up your nervous system along your spine passes through these points. When you are the tester, holding the points on yourself breaks the energetic circuitry with the other person and brings you back into your own energy loop, disengaging you and your beliefs from the other person. When you are being tested and place your hands on these points, it disengages you from your own thoughts about the matter and seems to allow the test to reveal a deeper level of information. Keeping your own beliefs and expectations from interfering is the most important single step you can take to ensure an accurate energy test.

3. **Focus Your Intention.** Your intention can influence a test's outcome. Rather than having to work around the effects of intention, however, you can use the fact that expectations influence energy testing, and resolutely set your intention for valid results. To demonstrate, find a food or other substance that tests strong on a friend. Repeat the test on the same substance without telling your friend what you are up to. On the second test, decide in advance that the test will show weak. Do not pull harder. Just see if this shift in your intention influences the results. It often does. So set your intention for an accurate test.

4. **Establish a Resonance between Tester and Tested.** You can create an energetic resonance with your partner:

a. Both take a deep breath in tandem.

b. Together release the breath along with all thoughts.

c. Test once the breath is released.

5. **Stay Alert.** Be sure the person is ready before beginning the test. Pull the arm smoothly. Apply pressure only long enough to determine whether the arm's resistance holds (generally between one and two seconds). If a muscle is locked in place or gives a fraction of an inch and then bounces back,

the test has shown that the energy is flowing through the muscle. Do not overwhelm the muscle "just to be sure." Along with learning the mechanics of energy testing, I am also interested in helping you cultivate a strong and reliable intuition. Energy testing is a superb means to that end because it provides tangible information about subtle energies. With each energy test you perform, you receive feedback that attunes your intuition to the flow of energies within you and around you as well as to the subtleties of energy testing itself.

6. **Practice under Double-Blind Conditions.** You can develop confidence in your ability to do an accurate energy test by practicing under conditions where you will get immediate feedback. You will need two other people. One will supervise the test, and one will be tested. Find some substances that the person being tested *knows* are healthy and agreeable to his or her system, such as organic apples or mint tea. Take other substances that you *know* would be bad to ingest, such as a bottle of ammonia. The person supervising the experiment takes a substance and puts it into the energy field of the person being tested in a way that neither you as the energy tester nor the person being tested can see, perhaps against the back of the person's shirt. Energy test. Do not be surprised if you do not get a perfect score at first, while you are learning the subtleties of the process. The double-blind test can be a great forum for learning the intricacies of energy testing, and it provides a benchmark as you continue to build your proficiency.

Energy testing is a method for assessing the body's energies and for evaluating how the environment, including the food you eat, is affecting them. You can use it to augment and refine the instructions provided throughout this book.

Notes

Introduction
EMBRACING YOUR EXQUISITE VIBRATION

1. "String theory" is a recent though controversial idea in physics that attempts to reconcile the "ill-concealed skeleton in the closet of physics" that what scientists understand about the behavior of the universe on extremely large scales (general relativity theory) and on extremely small scales (quantum mechanics) cannot both be right. If confirmed, string theory would resolve this contradiction and also provide a way of understanding how the material world is simultaneously both energy and matter. A good introduction to string theory is Brian Greene's *The Elegant Universe: Superstrings, Hidden Dimensions, and the Quest for the Ultimate Theory* (New York: W.W. Norton, 2003).

2. James Oschman, *Energy Medicine: The Scientific Basis* (New York: Harcourt, 2000).

3. Rephrased from Shakespeare, *Hamlet*, Act 2, Scene II.

4. Gary Null, Carolyn Dean, Martin Feldman, Debora Rasio, and Dorothy Smith, *Death by Medicine* (New York: Nutrition Institute of America, 2003).

Chapter 1

A MEDICINE CALLED ENERGY

1. Leopold Dorfer et al., "A Medical Report from the Stone Age?" *Lancet,* 1999; 354:1023–1025.
2. National Center for Complementary and Alternative Medicine. (2002). "What Is Complementary and Alternative Medicine?" Bethesda, MD: NCCAM. Retrieved December 3, 2006, from http://www.nccam.nih.gov/health/whatiscam.
3. Donna Eden and David Feinstein, "Energy Medicine: Uses in Medical Settings," 2006, paper posted at http://www.energymed.org/hbank/handouts/ener-med_in_medical_set.htm
4. Dawson Church, *The Genie in Your Genes: Epigenetic Medicine and the New Biology of Intention* (Santa Rosa, CA: Elite, 2007), pp. 137–138.
5. Church, *The Genie in Your Genes.*
6. Richard Gerber, *Vibrational Medicine,* 3rd ed. (Rochester, VT: Bear & Co.), p. 428.
7. William A. Tiller, *Psychoenergetic Science* (Walnut Creek, CA: Pavior, 2007).
8. Valerie Hunt, *Infinite Mind: The Science of Human Vibrations* (Malibu, CA: Malibu Publishing, 1995).
9. David Feinstein and Donna Eden, "Six Pillars of Energy Medicine," *Alternative Therapies in Health and Medicine* 2007, 14, 44–54. Available online from http://www.EnergyMedicinePrinciples.com.
10. Beverly Rubik, "The Biofield Hypothesis: Its Biophysical Basis and Role in Medicine," *Journal of Alternative and Complementary Medicine* 2002, 8, 703–717.
11. Valerie Hunt, *Infinite Mind.*
12. Timothy Ferris, *Coming of Age in the Milky Way* (New York: William Morrow, 1998).
13. Bill Bryson, *A Short History of Nearly Everything* (New York: Broadway, 2003), p. 141.
14. The Human Genome Project expected to identify some 120,000 genes but found, to everyone's astonishment, that fewer than 24,000 do the job (a simple roundworm contains 20,000 genes, so you can imagine the surprise).
15. Robert O. Becker, *Cross Currents: The Perils of Electropollution* (New York: Tarcher, 1990).
16. Lynn McTaggart, *The Field: The Quest for the Secret Force of the Universe* (New York: Harper, 2003), p. 45.
17. Cited in Bruce Lipton, "Mind Over Genes: The New Biology." Available from http://www.brucelipton.com/article/mind-over-genes-the-new-biology.
18. Harold S. Burr, *The Fields of Life* (New York: Ballantine, 1972).
19. Abraham R. Liboff, "Toward an Electromagnetic Paradigm for Biology and Medicine," *Journal of Alternative and Complementary Medicine* 2004, 10(1), 41–47.
20. Patricia Ellen Winstead-Fry and Jean Kijek, "An Integrative Review and Meta-Analysis of Therapeutic Touch Research," *Alternative Therapies in Health and Medicine,* 1999, 5(6), 58–67.

21. Melinda H. Conner, Genevieve Tau, and Gary E. Schwartz, "Oscillation of Amplitude as Measured by an Extra Low Magnetic Field Meter as a Physical Measure of Intentionality. Poster presentation, Toward a Science of Consciousness," University of Arizona, Tucson, AZ, May 2006.

22. Rolland McCraty, "The Energetic Heart: Bioelectromagnetic Communication within and between People," in Paul J. Rosch and Marko S. Markovs (eds.), *Clinical Applications of Bioelectromagnetic Medicine* (New York: Marcel Dekker, 2004), 541–562.

23. Reported in Paul Pearsall, *The Heart's Code* (New York: Broadway Books, 1998), p. 7.

24. James L. Ochsman, *Energy Medicine: The Scientific Basis* (London: Churchill Livingstone, 2000).

25. Bruce H. Lipton, *The Biology of Belief* (Santa Rosa, CA: Elite, 2005).

26. Gary Null, Carolyn Dean, Martin Feldman, Debora Rasio, and Dorothy Smith, *Death by Medicine* (New York: Nutrition Institute of America, 2003).

27. Ibid. p. 33.

28. Lipton, *The Biology of Belief,* p. 112.

29. Ibid., p. 115.

30. Ibid., p. 119.

Chapter 2
ENERGY TECHNIQUES FOR HEALTH AND VITALITY

1. Sections of this chapter are adapted from Chapters 3, 6, and 9 of *Energy Medicine.*

2. Certain points are called "forbidden points" in traditional acupressure, and activating the Large Intestine 4 point in a pregnant woman is not recommended. Forbidden points during pregnancy are discussed further in note 1 of Chapter 4.

3. Robert Frost, *Applied Kinesiology: A Training Manual and Reference Book of Basic Principles and Practices* (Berkeley, CA: North Atlantic Books, 2002).

4. Marcello Caso, "Evaluation of Chapman's Neurolymphatic Reflexes via Applied Kinesiology: A Case Report of Low Back Pain and Congenital Intestinal Abnormality." *Journal of Manipulative and Physiological Therapeutics* 2004, 27(1), 66–72.

5. Daniel G. Amen, *Making a Good Brain Great: The Amen Clinic Program for Achieving and Sustaining Optimal Mental Performance* (New York: Three Rivers Press, 2006).

6. Ibid., pp. 131–133.

7. Florence Peterson Kendall (ed.), *Muscles: Testing and Function, with Posture and Pain,* Fifth Edition (Baltimore: Lippincott, Williams & Wilkins, 2005).

8. Scott C. Cuthbert and George J. Goodheart, "On the Reliability and Validity of Manual Muscle Testing: A Literature Review," *Chiropractic & Osteopathy* (online journal) 2007, *15,* 4.

9. Henry Pollard, Bronwyn Lakay, Frances Tucker, Brett Watson, and Peter Bablis, "Interex-

aminer Reliability of the Deltoid and Psoas Muscle Test," *Journal of Manipulative and Physiological Therapeutics* 2005, 28(1), 52–56.

10. Educational applications of these methods are known as "Brain Gym." For further information, visit www.braingym.com and www.braingym.org.

11. David Feinstein, "Energy Psychology: A Review of the Preliminary Evidence," *Psychotherapy: Theory, Research, Practice, Training* (in press). Available from www.EnergyPsychologyResearch.com.

12. David Feinstein, Donna Eden, and Gary Craig, *The Promise of Energy Psychology: Revolutionary Tools for Dramatic Personal Change* (New York: Tarcher/Penguin, 2005). www.EnergyPsychEd.com.

Chapter 3
DANCING WITH YOUR HORMONES

1. Barry Sears, *The Age-Free Zone* (New York: HarperCollins, 2000).

2. J. D. Ratcliff, "The Endocrine Glands: Centers of Control," *Our Human Body* (Pleasantville, NY: The Reader's Digest Association, 1962), 271–276.

3. Norman Doidge, *The Brain That Changes Itself* (New York: Viking, 2007).

4. Mona Lisa Shulz, *The New Feminine Brain: How Women Can Develop Their Inner Strengths, Genius, and Intuition* (New York: Free Press, 1995).

5. Robert A. Wilson, *Feminine Forever* (New York: Evans, 1966).

6. See Judith A. Houck's "'What Do These Women Want?': Feminist Responses to *Feminine Forever*, 1963–1980" (*Bulletin of the History of Medicine,* 2003, 77:103–132) for a superb discussion of the juxtaposition and aftermath of Wilson's original article and Betty Friedan's *The Feminine Mystique,* both published in 1963.

7. Stephen Smith, "Hormone Therapy's Rise and Fall: Science Lost Its Way, and Women Lost Out" (*The Boston Globe,* July 20, 2003).

8. Leslie Kenton, *Passage to Power: Natural Menopause Revolution* (Carlsbad, CA: Hay House, 1998), p. 4.

9. Ibid.

10. The Boston Women's Health Book Collective, *Our Bodies, Ourselves: Menopause* (New York: Simon & Schuster, 2006), pp. 24–25.

11. Naomi Wolf, *The Beauty Myth* (Toronto: Vintage Books, 2007).

12. Nancy Etcoff, *Survival of the Prettiest: The Science of Beauty* (New York: Anchor, 2000).

13. Susun Weed, *New Menopausal Years: The Wise Woman Way* (Woodstock, NY: Ash Tree, 2002), p. 104.

14. Wolf, *The Beauty Myth.*

15. Christiane Northrup, *The Wisdom of Menopause: Creating Physical and Emotional Health and Healing During the Change* (New York: Bantam, 2001), p. 135.

16. Merlin Stone, *When God Was a Woman* (New York: Harcourt, 1978).

17. Judith Duerk, *Circle of Stones: Woman's Journey to Herself* (New York: New World Library, 2004).
18. Diana Schwarzbein, *The Schwarzbein Principle II: The Transition* (Deerfield Beach: Health Communications, 2002), p. 54.
19. James Wilson, *Adrenal Fatigue: The 21st Century Stress Syndrome* (Petaluma, CA: Smart Publications, 2002).

Chapter 4
RECLAIMING THE WISDOM OF YOUR MENSTRUAL TIME

1. Acupuncture points may be activated with needles, moxibustion (the burning of the herb *artemesia vulgaris* at the acupoint), electrical impulses, lasers, and manual stimulation such as massaging, tapping, or holding the points. I find acupuncture needles, moxibustion, and the electronic and laser stimulation of acupuncture points to be relatively intrusive compared with manual stimulation, and not necessarily more effective, though each method has its place. In the practice of acupuncture, certain points—including Spleen 9, Kidney 1, and Large Intestine 4—are not to be penetrated with a needle on a woman who is pregnant. My experience, and the experiences of many other hands-on energy medicine practitioners, has been that holding these points when sedating or strengthening a meridian does not cause adverse reactions. In fact, points that are considered unsafe to stimulate for pregnant women with the use of needles or moxibustion have, in clinical practice, been found to be safe when activated by electronic or laser stimulation. While I have no direct experience in using electronic or laser stimulation with pregnant women, these reports seem to support my experience that the even less invasive, hands-only methods taught in this book are safe. Nonetheless, since this is a self-help book and a practitioner is not there to monitor the procedures, the amount of time for holding the Kidney 1 point if you are pregnant has been, as a precautionary measure, reduced from two minutes to thirty seconds, long enough to get the process started but so short as to have no chance of causing an adverse effect. For a discussion of clinical findings related to acupuncture points that may have adverse effects, see Dr. John Amaro's "The 'Forbidden Points' of Acupuncture" at http://www.chiroweb.com/archives/18/10/01.html.

Chapter 5
SEXUALITY, FERTILITY, PREGNANCY, AND BIRTH

1. Michael Gurian, *What Could He Be Thinking: How a Man's Mind Really Works* (New York: St. Martin's Griffin, 2003).
2. Ibid.
3. Ellen Eatough's Web site and free "Love Tips" expands upon these and many other important ideas about sexuality. http://www.extatica.com.

4. Ibid.

5. Mary Eve, "Remembering V-Day," http://www.truthout.org/docs_2006/021407H.shtml.

6. Margot Anand, *The Art of Sexual Ecstasy* (New York: Tarcher/Penguin, 1989).

7. Kenneth Ray Stubbs, *The Essential Tantra: A Modern Guide to Sacred Sexuality* (New York: Tarcher/Putnam, 1999).

8. David Feinstein, Donna Eden, and Gary Craig, *The Promise of Energy Psychology: Revolutionary Tools for Dramatic Personal Change* (New York: Tarcher/Penguin, 2005).

9. Alan Batchelder was the therapist.

10. "The Energies of Love" has been one of our most popular classes. An introduction to this topic is on a DVD of that name, available from www.innersource.net

11. Christine Gorman, "The Limits of Science," *Time,* April 15, 2002, 159 (15), p. 52.

12. Randine Lewis, *The Infertility Cure* (New York: Little, Brown & Co., 2004), p. 15.

13. Jorge E. Chavarro, Walter C. Willett, and Patrick J. Skerrett, *The Fertility Diet* (New York: McGraw-Hill, 2008).

14. Ibid., pp. 12–13.

15. Ibid., pp. 149–151.

16. Ibid.

17. Ibid., p. 282.

18. Vicki Hufnagel, *No More Hysterectomies,* rev. ed. (New York: Penguin, 1989).

19. Ibid., p. 60.

Chapter 6
MENOPAUSE—GATEWAY TO YOUR *SECOND PRIME* OF LIFE

1. To examine your life as a mythic journey, see David's award-winning *The Mythic Path: Discovering the Guiding Stories of Your Past: Creating a Vision for Your Future* (Book of the Year—2007 U.S. Book News Awards, Psychology/Mental Health Category). For an overview, visit www.innersource.net.

2. Pat Wingert and Barbara Kantrowitz, *Is It Hot in Here? Or Is It Just Me? The Complete Guide to Menopause* (New York: Workman Publishing Company, 2006).

3. Leslie Kenton's *Passage to Power: Natural Menopause Revolution* (Carlsbad, CA: Hay House, 1995) is a wonderful synthesis of the biological and spiritual dimensions of menopause.

4. Susun Weed's *New Menopausal Years: The Wise Woman Way* (Woodstock, NY: Ash Tree, 2002) is a distillation of what *works* after having interviewed some 50,000 menopausal women over a 13-year period.

5. Christiane Northrup's *The Wisdom of Menopause: Creating Physical and Emotional Health and Healing During the Change* (New York: Bantam, 2001) is a gynecologist's highly personal yet information-packed guide for menopausal women.

6. John Lee's *What Your Doctor May Not Tell You about Menopause: The Breakthrough Book on* Natural *Hormone Balance*, rev. ed. (New York: Warner, 2004), is an analysis of the good science and the bad science behind hormone replacement therapy by a gynecologist who was a pioneer in the use of natural hormones.

7. Susan Lark's *Hormone Revolution: Yes, You Can Naturally Restore & Balance Your Own Hormones* (Los Altos, CA: Portola Press, 2008) is rich with information and strategies for hormone balance, menopause, perimenopause, fatigue, hypertension, and many other areas that affect your health and well-being. Based on Dr. Lark's 30 years of medical practice and innovative teaching, it focuses on five hormones—estrogen, progesterone, testosterone, pregnenolone, and DHEA—and includes a section on energy medicine.

8. Kenton, *Passage to Power*, p. 205.

9. Leonard Shlain, *Sex, Time and Power: How Women's Sexuality Shaped Human Evolution* (New York: Penguin, 2004), p. 183.

10. Ibid., p. 159.

11. Ibid., p. 184.

12. Ibid., p. 95.

13. Ibid.

14. Ibid., p. 96.

15. Jacques E. Rossouw, et al., "Postmenopausal Hormone Therapy and Risk of Cardiovascular Disease by Age and Years Since Menopause." *JAMA* 2007, 297(13), 1,465–1,477.

16. Lee, *What Your Doctor May Not Tell You About Menopause*.

17. Ibid., pp. 3–4.

18. Germaine Greer, *The Change: Women, Aging and Menopause* (New York: Fawcett Columbine, 1991).

19. Sandra Coney, *The Menopause Industry* (Alameda, CA: Hunter House, 1994), p. 25.

20. Cited in Lee, *What Your Doctor May Not Tell You About Menopause*, p. 52.

21. Ibid., p. 57.

22. Bethany Hays, "Solving the Puzzle of Hormone Replacement," *Alternative Therapies in Health and Medicine* 2007, 13(3), 50–57.

23. Christiane Northrup, *The Wisdom of Menopause*, PBS Television Special, 2001.

24. Susan Lark, *A Woman's Guide to Embracing "The Change," Special Report of The Lark Letter* (Forrester Center, WV: Healthy Directions: 2005).

25. Diana Schwarzbein, *The Schwarzbein Principle II: The Transition—A Regeneration Program to Prevent and Reverse Accelerated Aging* (Deerfield Beach, FL: HIC, 2002), pp. 44–46.

26. Leslie Kenton, *Passage to Power*, p. 105.

27. Ibid., p. 110.

Appendix

TALKING WITH YOUR BODY'S ENERGIES—THE ART OF ENERGY TESTING

1. Adapted from Chapter 2 of *Energy Medicine* (2nd ed., New York: Tarcher/Penguin, 2008).

2. Dean I. Radin, "A Possible Proximity Effect on Human Grip Strength," *Perceptual and Motor Skills* 1984 (58), 887–888.

3. Personal communication, January 16, 1998.

4. Scott C. Cuthbert and George J. Goodheart, "On the Reliability and Validity of Manual Muscle Testing: A Literature Review." *Chiropractic & Osteopathy* (online journal) 2007, 15:4.

5. William Caruso and Gerald Leisman, "A Force/Displacement Analysis of Muscle Testing," *Perceptual and Motor Skills* 2000, 91, 683–692.

6. D. Monti, J. Sinnott, M. Marchese, E. Kunkel, and J. Greeson, "Muscle Test Comparisons of Congruent and Incongruent Self-Referential Statements." *Perceptual and Motor Skills* 1999, 88, 1019–1028.

7. Arden Lawson and Lawrence Caleron, "Interexaminer Agreement for Applied Kinesiology Manual Muscle Testing," *Perceptual and Motor Skills* 1997, 84, 539–546.

8. Chang-Yu Hsieh and Reed B. Phillips, "Reliability of Manual Muscle Testing with a Computerized Dynamometer," *Journal of Manipulative and Physiological Therapeutics* 1990, 13, 72–82.

9. Gerald Leisman, Robert Zenhausern, Avery Ferentz, Tesfaye Tefera, and Alexander Zemcov, "Electromyographic Effects of Fatigue and Task Repetition on the Validity of Estimates of Strong and Weak Muscles in Applied Kinesiological Muscle-Testing Procedures," *Perceptual and Motor Skills* 1995, 80, 963–977.

10. Gerald Leisman, Philip Shambaugh, and Avery Ferentz, "Somatosensory Evoked Potential Changes During Muscle Testing," *International Journal of Neuroscience* 1989 (45), 143–151.

11. Stephanie Eldringhoff and Victoria H. Matthews, "Frozen and Irregular Energies: Hidden Energy Stumbling Blocks," 2006, http://www.energymed.org/hbank/handouts/frozen_irregular_energies.htm.

Resources

Most people have no idea of the giant capacity we can immediately command when we focus all of our resources on mastering a single area of our lives.

—Tony Robbins

Join or Start a Local Study Group

A great way to learn energy medicine is to find study buddies with whom you can practice. This may be one other person or a small group, and it may be as simple as discussing this book or practicing techniques together after watching them on a DVD. Many communities have local energy medicine study groups. You can find them, along with other individuals interested in energy medicine, at www.Energy MedicineDirectory.com.

Find an Energy Medicine Practitioner

Every local community is enjoying a rapid increase in the number of health practitioners who incorporate an energy medicine perspective. Practitioners may be found in all of the healing professions, from physicians and chiropractors to nurses to personal trainers to massage therapists. Practitioners of acupuncture, qi gong, Reiki, Ayurveda, Applied Kinesiology, homeopathy, Touch for Health, Healing Touch, and Therapeutic Touch, among many others, are all working directly with the body's subtle energies. An excellent guide to finding a qualified energy practitioner in your own local community, as well as a listing of practitioners trained in my approach to energy medicine, can be found in the "Practitioners and Links" area at www.innersource.net.

Find Classes and Trainings

Local communities are continually offering classes in various aspects of energy medicine. In larger cities, widely distributed free monthly local magazines announce dozens every month. Courses with David or me can be found on the "Classes" area at www.innersource.net.

Eden Energy Medicine Certification Program

Taught by gifted practitioners who have been studying with me for many years, and who are still overseen by me, this two-year program has earned rave reviews from hundreds of graduates. Energy medicine is a career for the future, and this program is a route into it. Learn more on the "Certification Program" area at www.innersource.net.

Additional Books, Videos, and Other Learning Resources

Books by David and/or me have collectively won four national awards, including two Book of the Year designations in their respective categories. The six-hour *Energy Medicine: The Essential Techniques* video program takes most of the exercises from *Energy Medicine* and shows me personally instructing you in how to use them. It is like bringing me into your living room. See all of our books, DVDs, and CDs, and the superb Energy Medicine Kit by Sounds True, at www.innersource.net.

Home Study Resources

Many of our books and videos can form the basis of home study programs with exams, certificates of completion, and professional continuing education credit available. Learn more at www.EnergyHomeStudy.com.

The Energy e-Letter

Learn about numerous topics of interest; new books, DVDs, and other resources; and upcoming classes and training events, and generally stay informed about

David's and my approach to energy healing. You can sign up free from nearly any page at www.innersource.net. Your name will never be given to others.

The Energy Medicine Handout Bank

As more and more of my students have been teaching energy medicine, they have put great amounts of energy into developing handouts for their classes. They have often developed material that was similar to handouts used by others. We decided to take the best of these handouts and place them on our "Handout Bank." The Handout Bank is a free resource designed to (1) help make energy medicine more widely accessible, (2) aid those who are teaching classes or providing services in energy medicine, and (3) create a high-quality archive of principles and methods. It is designed for the energy medicine practitioner, but others interested in the field may also find it a valuable resource. The Handout Bank is posted on the site of our sister organization, the nonprofit Energy Medicine Institute, www.EnergyMed.org.

Getting an Energy Medicine Perspective on Health Questions

Energy medicine does not *diagnose* or *treat* illness or disease. Instead, it corrects energetic imbalances that are at the foundation of health and vibrancy. But physical symptoms often provide clues about the types of energy imbalances the body needs to have addressed. I have received thousands of inquiries about how to apply energy medicine with various health-related concerns. While the numbers now make it impossible to respond to people personally, for many years I did just that, so I have hundreds of responses to such questions. Our staff has selected questions and answers that may apply to others with similar concerns, concealed the writers' identities, and edited and posted them so the information might serve many. These are well indexed in the "Questions and Answers" section at www .innersource.net.

Learn About Energy Psychology

Applying the principles of energy medicine to emotional problems and to promoting peak vitality is proving to be one of the most exciting developments in the field of psychology. Learn more at www.EnergyPsychEd.com.

Index

Other Books and Programs
of Interest

The *Energy Medicine for Women* DVD program, a highly recommended companion to this book, shows Donna Eden demonstrating most of the procedures the book presents.

Energy Medicine by Donna Eden (with David Feinstein, Ph.D.), the bestselling classic, is available now in a thoroughly revised and up-to-date tenth anniversary edition. It also has a companion DVD program, *Energy Medicine: The Essential Techniques,* which features six hours of Donna demonstrating the most fundamental procedures of energy medicine.

An energizing first look at energy medicine is *Introduction to Energy Medicine,* a DVD based on a two-hour live seminar with Donna. Another introductory resource, the *Energy Medicine Kit,* was a One Spirit featured selection. A variety of other basic to advanced DVD programs is also available for bringing energy medicine into your living room.

The Promise of Energy Psychology, by David Feinstein, Ph.D., Donna Eden, and Gary Craig, provides simple step-by-step instructions for creating constructive changes in your energy system *and brain chemistry.* These changes can help you alter behavioral patterns; overcome psychological challenges such as inappropriate fear, guilt, shame, jealousy, and anger; and succeed with challenging goals. A companion DVD program, *Introduction to Energy Psychology,* explains the essential concepts, and demonstrates the basic procedures emerging from this revolutionary development within psychology. The book was a finalist in the Self-Help category at the 2007 PubInsider National Book Awards (the Indies).

Personal Mythology: Using Ritual, Dreams, and Imagination to Discover Your Inner Story and a companion CD present a 12-week program for transforming the deep guiding myths that shape our lives. Its new third edition, which adds an energy psychology module, was named a *USA Book News'* 2007 Book of the Year (Psychology/Mental Health category) and 2007 Book of the Year (New Age Nonfiction category) at the Indies Awards. The book provides a compass for these troubling times.

Live classes, ranging from introductory workshops to a two-year professional certification program in energy medicine, are available. A catalog describing other materials and training programs will be sent upon request (contact energy@ innersource.net). Professional continuing education credit is available for many of these programs.

www.innersource.net

24-hour Order Line: 800-835-8332

Innersource, 777 East Main Street, Ashland, OR 97520

About the Authors

DONNA EDEN is among the world's most sought, most joyous, and most authoritative spokespersons for Energy Medicine. Her abilities as a healer are legendary: she is able to accurately determine the energetic causes of physical and psychological problems and to devise highly effective interventions. Since childhood, Eden has been able to see the flow of the body's energies, and from this clairvoyant ability she has developed a system for teaching others, who do not have this gift, to enhance their health and vitality by working with their energies. More than 50,000 people from all over the world have learned how to reclaim their natural healing capabilities in her classes and workshops. Her bestselling book *Energy Medicine* has been translated into more than a dozen languages and is a classic in its field. According to Caroline Myss: "The contribution Donna Eden has made with *Energy Medicine* will stand as one of the backbone studies as we lay a sound foundation for the field of holistic medicine." Visit Donna Eden's website at www.LearnEnergy Medicine.com.

DAVID FEINSTEIN, PH.D., a clinical psychologist, has served on the faculties of the Johns Hopkins University School of Medicine and Antioch College. Among his major works are *The Promise of Energy Psychology*, *Rituals for Living and Dying*, and *Personal Mythology*. He has contributed more than fifty articles to the professional literature, and his books have won four national awards, including the *USA Book News* 2007 Book of the Year Award in the psychology/mental health category (for *The Mythic Path*). Visit EnergyPsychEd.com.